Water and Disasters

This book is the first major attempt to address, comprehensively and in-depth, the many issues associated with water and disasters. It is particularly relevant and topical in view of the increasing frequency and intensity of water-triggered disasters that have afflicted the world in recent years, among them the Indian Ocean Tsunami and Hurricane Katrina.

Water and Disasters is a global survey—and assessment of the causes, consequences and post-recovery policies—concerning water disasters. The chapters include empirical studies, case histories, conceptual-theoretical investigations, policy perspectives, institutional analysis, and risk analysis, among others. The book features a comprehensive discussion of the 2004 Indian Ocean Tsunami, as well as major floods and droughts in England, Wales, China and the western United States. It also includes chapters on advances in decision support systems for flood disaster management and rainfall insurance.

This volume should be of special interest to disaster management planners and practitioners globally, primarily in the domain of water, in crafting creative solutions for tackling the disasters effectively, efficiently and rapidly.

This book was previously published as a special issue of *International Journal of Water Resources Development.*

Chennat Gopalakrishnan is Professor of Natural Resource Economics at the University of Hawaii. He has published five books and approximately ninety journal articles and technical papers on current and emerging issues in natural resources economics and policy. Gopalakrishnan has directed a number of research projects dealing with water institutions, water markets and pricing, and water conservation and management. Gopalakrishnan is the recipient of numerous awards and honours, including Fellow of the American Water Resources Association.

Norio Okada is Professor of Social Systems Engineering at Kyoto University, Japan. He has directed a number of research projects in the area of integrated disaster risk management, with special reference to Japan. Dr. Okada has chaired or served on numerous scientific advisory committees for international conferences on integrated disaster risk management.

Water and Disasters

Edited by
Chennat Gopalakrishnan and Norio Okada

Routledge
Taylor & Francis Group
LONDON AND NEW YORK

First published 2007 by Routledge
2 Park Square, Milton Park, Abingdon, Oxon, OX14 4RN

Simultaneously published in the USA and Canada
by Routledge
270 Madison Avenue, New York, NY 10016

Routledge is an imprint of the Taylor & Francis Group, an informa business

Typeset in Times 10/12pt by the Alden Group, Oxfordshire
Printed and bound in Great Britain by MPG Books Ltd, Bodmin, Cornwall

British Library Cataloguing in Publication Data
A catalogue record for this book is available from the British Library

ISBN 10: 0-415-45426-3 (hbk)
ISBN 10: 0-415-46207-X (pbk)

ISBN 13: 978-0-415-45426-1 (hbk)
ISBN 13: 978-0-415-46207-5 (pbk)

CONTENTS

List of Contributors

P.D. Bates, Bristol University, UK

R.J. Dawson, University of Newcastle upon Tyne, UK

Lindsey J. Ellingson, Colorado State University, Fort Collins, USA

W. Marshall Frasier, Colorado State University, Fort Collins, USA

Yi Ge, Key Laboratory of Environmental Change and Natural Disasters, Beijing, China

Chennat Gopalakrishnan, University of Hawaii at Manoa, USA

Weiping Guo, Key Laboratory of Environmental Change and Natural Disasters, Beijing, China

J.W. Hall, University of Newcastle upon Tyne, UK

Clare L. Johnson, Middlesex University, UK

Ron Lacewell, Texas A&M University, College Station, USA

Jason K. Levy, Western Washington University, USA

Zhaohui Lin, Chinese Academy of Sciences, China

Ari Michelsen, Texas A&M University, El Paso, USA

R.J. Nicholls, Southhampton University, UK

Norio Okada, Kyoto University, Japan

Edmund C. Penning-Rowsell, Middlesex University, UK

Eric C. Schuck, Linfield College, Oregon, USA

Zhuping Sheng, Texas A&M University, El Paso, USA

Peijun Shi, Beijing Normal University, China

Jerry R. Skees, University of Kentucky, Lexington, USA

Allen Sturdivant, Texas A&M University, College Station, USA

Sylvia M. Tunstall, Middlesex University, UK

Wendy J. Umberger, Colorado State University, Fort Collins, USA

Robert S. Webb, NOAA Climate Diagnostics Center and Western Water Assessment, Boulder, Colorado, USA

Yi Yuan, National Disaster Reduction Center of China, Beijing, China

Kimberly A. Zeuli, University of Wisconsin-Madison, USA

Water and Disasters—Crafting Creative Solutions

An aspect of water resources management that has received relatively little attention in the discourse on water pertains to water and disasters. The destructive potential of water has been well recorded in history going as far back as biblical times. Recent years have witnessed instance after instance of the catastrophic consequences of floods and droughts throughout the world. Particularly noteworthy in this context are the great Indian Ocean tsunami of 26 December 2004, the major Mumbai floods in May 2005, and, in August 2005, the catastrophic hurricane Katrina and the resulting devastating floods in New Orleans. In terms of sheer destruction of physical infrastructure there are few natural disasters on record that match Katrina. Preliminary estimates put economic losses from the hurricane in excess of US$200 billion (with hundreds of people reportedly dead), even surpassing the damage inflicted by the Indian Ocean tsunami. Equally destructive are major droughts that have plagued different parts of the world, especially in the African continent, throughout this century causing colossal losses and suffering. Devoted to the theme of 'Water and Disasters', this volume is a major effort to examine in depth and comprehensively the multiple facets of water in the context of disasters.

Several issues central to developing and implementing an effective, efficient, and viable integrated disaster management plan have been identified by expert analysts in the field. In the wake of the killer hurricane Katrina and other major natural disasters, it is important to pay special attention to these suggestions. These include: (1) the need for an effective early warning system that can rapidly disseminate to all concerned parties the details of an imminent disaster; (2) the need for a co-ordinated plan at the local, state, and national levels to implement at short notice the search, rescue, and evacuation of all affected residents; (3) the need for a clear delineation of responsibilities among many different agencies without compromising the ability for instant co-ordination at all levels; (4) the case for a state-of-the-art system for long-term forecasting of flood and drought disasters; (5) the case for effective public–private partnership in disaster management, mitigation, and relief efforts; (6) the ability for rapid economic assessment of flood and drought damage; (7) synchronization of relief efforts when floods or droughts strike two or more adjacent countries simultaneously; and (8) plans in place for the effective deployment of international aid in the wake of major disasters.

Robust institutions for integrated disaster risk management are absolutely essential to implement in a timely fashion disaster warning, evacuation, emergency relief, mitigation, and post-recovery efforts. The recent Katrina debacle reinforces the concerns that have been articulated in the discourse on integrated disaster risk management about institutional impediments. For instance, the excruciatingly slow delivery of much-needed assistance by the Federal Emergency Management Agency (FEMA) to the residents of New Orleans and other Gulf Coast areas after the disaster struck has been singled out by many observers and analysts as the crucial contributing factor to the widespread devastation that followed. Likewise, in the case of the great Indian Ocean tsunami, the absence of a tsunami warning system was largely responsible for the extensive destruction in the Indian Ocean countries.

A survey of historical floods and droughts points to the lack of a well-conceived and coherent disaster management policy as the major cause for the exacerbation of the enormous tragic consequences that followed. Gopalakrishnan & Okada (2003, 2004) have argued the case for a thorough overhaul of the institutional component of integrated disaster risk management as currently conceived. Their review of disaster management institutions has indicated significant weaknesses in the ability of the current system globally to effectively implement integrated disaster risk management. The recent Katrina debacle reinforces the concerns articulated in these papers. Gopalakrishnan & Okada (2003) have identified eight key elements for the design of dynamic new disaster management institutions. Based on their analysis the authors have suggested six approaches for incorporation in the design and development of new institutions for the successful implementation of integrated disaster risk management in its multiple phases. These are institutional integration, public–private partnership, information sharing and technology transfer, disaster diagnosis, legislative revamping, and cultural calibration.

The chapters corroborate several issues noted above and thus further strengthen the case for institutional revamping. The opening paper by Levy & Gopalakrishnan discusses and evaluates the socio-economic damage and human toll of the Sumatra–Andaman earthquake of 26 December 2004 and the resulting tsunami, the most devastating (property damage and human toll combined) in recorded history. The authors make the case for implementing a reliable tsunami warning and mitigation system for the Indian Ocean countries and the development of greater tsunami resilience in the region. Johnson *et al.* discuss four of the most critical flood disasters that have occurred in England and Wales in the past 50 years and assess their impact on flood policy formulation. Dawson *et al.* investigate and quantify the likely flood impacts of the Thames estuary under different plausible, but unlikely, sea level rise scenarios. The authors point out that future management strategies need to be adaptive and robust to effectively deal with climate change uncertainty. Levy *et al.* discuss the challenges and opportunities that face decision makers in incorporating and implementing advances in decision support systems for flood disaster management. The authors propose a flood decision support system architecture that capitalizes on the latest advances in flood risk management such as remote sensing, geographic information systems (GIS), hydrologic models, and numerical weather predictions, etc. Shi *et al.* present a comprehensive review of integrated risk management of flood disasters in the metropolitan areas of China. This paper elucidates the relationship and the interactions between urbanization in the metropolis and the process of flood disaster changes and proposes a solution to alleviate these fluctuations through the adjustment of the land use structure and pattern in metropolitan areas. Sheng *et al.* provide an analysis of rapid economic assessment of flood control failure along the Rio Grande via a comprehensive case study. The economic losses avoided by four major flood control projects on the Rio Grande are estimated using a novel approach combining satellite imagery, GIS, and economic methods. This is a particularly timely contribution in light of the singular lack of rapid action on the part of FEMA and other responsible parties in the wake of hurricane Katrina.

The last two papers change the focus from floods to droughts, a water disaster that can be as catastrophic as flood events, as the recent drought episodes in the African continent have demonstrated. Schuck *et al.*, using survey data from the worst drought in Colorado's history, point out how drought conditions influence the choice of irrigation systems by irrigators. They demonstrate with supportive evidence how the adoption of technically efficient irrigation systems can mitigate the effects of droughts by enabling irrigators to maintain water consumption with curtailed applications. In the final paper Zeuli & Skees cxamine the use of a rainfall index contract as a tool to improve drought management with the aid of a probing case study of an irrigation district in New South Wales, Australia.

The editors hope this volume will receive the attention of disaster management planners and practitioners globally, especially in the domain of water, and jolt them into crafting creative solutions to tackling thse disasters effectively, efficiently, and rapidly.

Chennat Gopalakrishnan
University of Hawaii at Manoa, Honolulu, HI, USA

Norio Okada
Kyoto University, Kyoto, Japan

References

Gopalakrishnan, C. & Okada, N. (2003) Enhancing the effectiveness of integrated management for disaster risk: the case for institutional innovations. Paper presented at Third International Symposium on Integrated Disaster Risk Management, Kyoto, 3–5 July.

Gopalakrishnan, C. & Okada, N. (2004) Reflections on implementation science, in: *Proceedings of the Ravello Forum on 'Integrated Disaster Risk Management'*, pp. 133–144 (Ravello: Disaster Prevention Research Institute, Kyoto University, Japan).

Promoting Disaster-resilient Communities: The Great Sumatra–Andaman Earthquake of 26 December 2004 and the Resulting Indian Ocean Tsunami

JASON K. LEVY & CHENNAT GOPALAKRISHNAN

Introduction

The Sumatra–Andaman earthquake occurred at 00:58:53 coordinated universal time (07:58:53 local time) on 26 December 2004 at a depth of 30 km (18.6 miles below sea level), just north of Simeulue Island off the western coast of northern Sumatra, Indonesia. This earthquake was the largest seismic event on Earth in the past 40 years and the largest ever recorded with global digital seismometry. The Sumatra–Andaman event was also the third largest earthquake in the past century behind the 1960 southern Chile and 1964 Prince William Sound (Alaska) seismic events. Lasting for nearly 10 min (most major earthquakes last no more than a few seconds), the 26 December 2004 Indian Ocean earthquake generated the most devastating tsunami in recorded history, although the final death toll may never be known exactly (many bodies were washed out to sea, buried by debris, etc.). More than 283 000 people have been killed or are reported missing and approximately 1.7 million people lost their homes, while another 2 million have probably been displaced by the disaster. More deaths are possible due to poor sanitation, although the unprecedented humanitarian response largely averted the threat of starvation and epidemics. The tsunami devastated coastal regions of 13 Indian Ocean countries, including the shores of Indonesia, Sri Lanka, South India, Thailand, and other countries. The most tsunami-affected nations (and the number of deaths by country) are shown in Figure 1. Damage was reported as far as the east coast of Africa, with the furthest tsunami death occurring at Port Elizabeth in South Africa, 8000 km (5000 miles) away from the earthquake's epicentre. In many locations, the waves reached as far as 2 km (1.24 miles) inland (Pearce & Holmes, 2005).

The total energy released by the earthquake and the resulting tsunami waves has been estimated as 4.3 EJ (as much energy used in the USA in 6 months) and 20 PJ, respectively (Bilham, 2005). Initial reports by the National Oceanic Atmospheric Administration's (NOAA's) Pacific Tsunami Warning Center (PTWC), the US Geological Survey (USGS), and others significantly underestimated the earthquake's magnitude (USGS, 2004). The most definitive estimate so far, in March 2005, estimated the earthquake's magnitude at 9.15 on the Richter scale (Hanson, 2005), although its magnitude has been estimated to be considerably higher (McKee, 2005). The hypocentre of the main earthquake was at 3.316° N, 95.854° E (3° 19′ N, 95° 51.24′ E), some 160 km (100 miles) west of Sumatra, at a depth of 30 km (18.6 miles) below mean sea level. The earthquake occurred at the extreme western end of the Ring of Fire, an earthquake belt that accounts for over 80% of the world's largest earthquakes.

Large earthquakes have occurred throughout history, but accurate measurements have only been available since the 20th century. The Sumatra–Andaman event, with its consequent tsunami, was the first great earthquake to be observed with a world-wide seismic network and global positioning system (GPS). Seismic, satellite, and GPS observation of this earthquake shows that the Sumatra–Andaman earthquake caused the longest fault rupture ever: over 1200 km (750 miles) of faultline slipped up about 20 m (65 feet). The unstable faultline triggered powerful aftershocks, primarily off the Andaman Islands, the Nicobar Islands and the region of the original epicentre. The largest aftershock was epicentred off the Sumatran island of Nias at 23:09:37 local time on 28 March 2005. At magnitude 8.7, this earthquake ranks as the seventh largest since 1900, when reliable measurements became available.

In early 2005, National Aeronautics and Space Administration scientists noted that the shift of mass and the massive release of energy disrupted the Earth's rotation and shortened the length of each day by 2.68 μs due to a decrease in the Earth's oblateness (Cook-Anderson & Beasley, 2005). In addition, the North Pole shifted by about 2.5 cm (1 inch) and the planet is now slightly less oblate (flattened at the poles). The earthquake was so powerful that it shifted

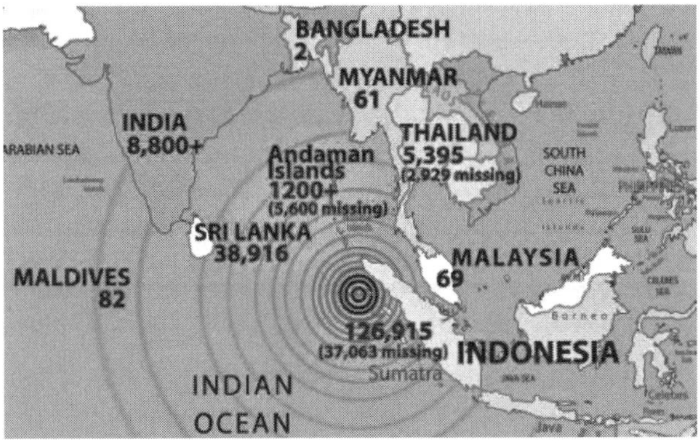

Figure 1. Indian Ocean nations most affected by the 26 December 2004 Indian Ocean tsunami. *Source:* Center of Excellence in Disaster Management (2005).

the axis around which the Earth revolves by approximately 2.5 cm (1 inch) in the direction of 145° E longitude (Schechner, 2004), and led to an oscillation of the Earth's surface of about 20–30 cm (8–12 inches), equivalent to the effect of the tidal forces caused by the Sun and Moon combined. The shock waves of the Sumatra–Andaman earthquake were felt as far away as the USA, where vertical movements of 3 mm (0.12 inches) were recorded. High-resolution multi-beam sonar showed that the earthquake had significantly altered sea bed topography, including a newly formed oceanic trench, several kilometres wide (Knight, 2005). Also, millions of blocks of rock were dragged up to 10 km (6.2 miles) across the sea bed by the force of the displaced water.

The Sumatra–Andaman earthquake sent a sea surge in the Indian Ocean that reached speeds of 270 m.p.h. (450 km/h) in open water. Sweeping across the Indian Ocean, the earthquake-triggered waves affected 13 Indian Ocean nations. As documented by Global Sea Level Observing System stations, it approached Sri Lanka after about 3 h. Within 4 h, the tsunami had reached Diego Garcia in the Indian Ocean and in less than 10 h it had reached Somalia and other countries along the eastern coast of Africa. Nearing land, the tsunami slowed to approximately 27m.p.h. (45 km/h), where it was pushed upwards, producing enormous waves that devastated coastal regions.

Indian Ocean Tsunami Background: Physical Science

A tsunami (derived from the Japanese word meaning 'harbour wave') is a series of rapidly moving, long-wavelength, long-period ocean waves generated by impulsive, large-scale disturbances of the Earth's crust (ocean floor). Tectonic processes are the most common mechanism of tsunami generation. The Earth's surface is composed of a small number of tectonic plates (after the Greek word *tekton*, or builder), large crustal slabs about 70–250 km (40–150 miles) thick that contain both the continents and the sea floor. Floating on a viscous underlayer of melted rock (the asthenosphere), the plates (and their superimposed continents and ocean basins) slowly grind past each other at rates of up to 10 cm/year (several inches/year). At subduction zones (where one crustal plate is pulled underneath the edge of an adjoining plate) frictional stresses build and release, creating mountains, earthquakes, and volcanic eruptions (see Figure 2).

Destructive tsunamis originate primarily in subduction zones around the Pacific Rim, typically generated by large earthquakes with a magnitude greater than 7.5. The India Plate is part of the great Indo-Australian Plate, which underlies the Indian Ocean and the Bay of Bengal, and is drifting north-east at an average of 6 cm/year (2 inches/year). The India Plate meets the Burma Plate (which is a section of the Eurasian Plate) at the Sunda Trench. At this point the India Plate subducts the Burma Plate, which carries the Nicobar Islands, the Andaman Islands and northern Sumatra. The 26 December 2004 Sumatra–Andaman earthquake ruptured the boundary between the Indo-Australian plate and the south-eastern portion of the Eurasian Plate causing the sea-bed to rise several metres, and displacing significant volumes of sea water. The resulting train of water waves radiated outward in all directions from the disturbance at great speeds.

While capable of causing large-scale destruction, the shifting of tectonic plates and other geological processes are also partly responsible for supporting life and ensuring the high habitability of the Earth: the slow, constant recycling of the planetary crust is essential for the evolution of complex life and planetary renewal. As declared by Frank Press, past president of the National Academy of Sciences, "On balance, it's possible that life on earth would not have originated without plate tectonics, or the atmosphere, or the oceans" (Broad, 2005). The motion of tectonic plates provides a number of important benefits for the planet which accumulate slowly and globally over millennia (Schlesinger, 1997; Press *et al.*, 2004). Tectonic processes began billions of years ago, helping to create life on Earth. Crustal recycling is believed to have created the Earth's primordial oceans and atmosphere as volcanoes released large quantities of water vapour, nitrogen, carbon dioxide, and other gases. In fact, the Earth's first organisms (ancestors of microbes known as thermophiles, which today thrive in hot regions deep underground) are believed to have arisen in the deep sea-bed, along hot volcanic gashes. Tectonic activity not only builds continents, mountain ranges, and island chains but also enriches soils (volcanic soils are among the highest-quality soils on the planet).

In particular, tsunamis can distribute rich sediments from river systems into lowlands, bringing fertile soils into the coastal plains. Also, water travelling through the sea-bed's hot volcanic channels concentrates metals (copper, gold, silver, etc.) into rich deposits that the churning Earth pushes onto dry land. Finally, subduction processes play a critical role in regulating global chemistry: oceans slowly pass through the tectonic system approximately once every million years. This increases nutrients in the biosphere and regulates important elements (boron, calcium, etc.). Plate motion also plays a vital role in recycling carbon dioxide, thus regulating the planet's temperature and creating a stable long-term climate (Figure 2).

It is often difficult to determine whether a particular tsunami has been caused directly by an earthquake or by a submarine landslide (triggered by the earthquake). Submarine landslide-generated tsunamis typically involve the rapid transport of rock or sediment and involve the movement of sub-sea floor material from one place to another (generating a

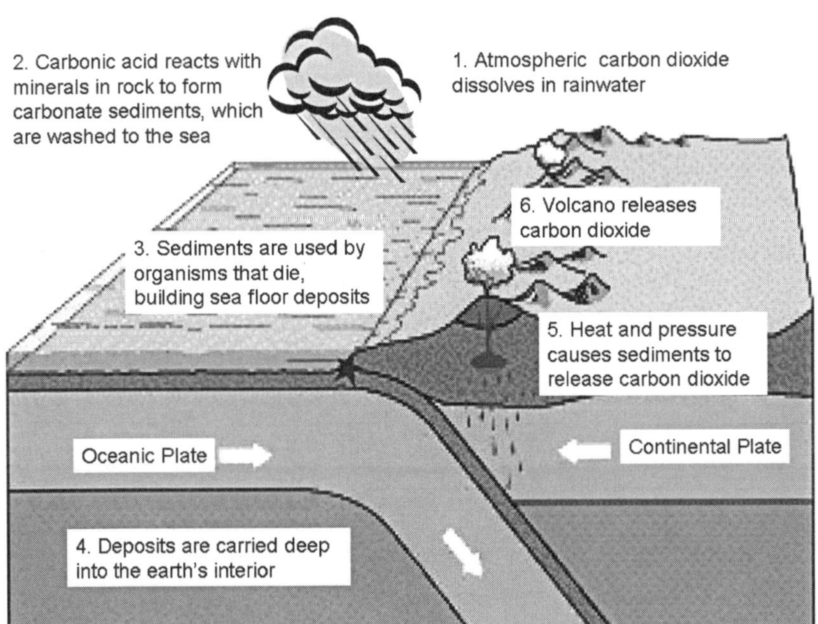

Figure 2. Earthquakes, volcanoes and plate tectonics. *Source:* Modified from Broad (2005).

reverse flow of water). Landslide-generated tsunami may also trigger volcanic eruptions, although volcanic eruptions can directly generate tsunamis. Catastrophic tsunamis occur when 'near Earth objects' (asteroids, meteors, comets, etc.) hit the Earth. Fortunately, such events are very rare, with a frequency of around once in every 100 000 years. However, 65 million years ago the mass extinction of 70% of all species living on Earth (the so-called K-T event) is believed to have been caused by the impact of a 10 km diameter asteroid hitting the Yucatan region of Mexico. Human activities (construction, explosions, etc.) can also generate tsunamis. They often occur in densely populated areas (ports, harbours, etc.) where economic and social damage is extreme. For example, on 3 November 1994 in Skagway, Alaska, a submarine landslide was triggered when 1 million m^3 of rubble and sediment were reported to have slid into the fjord during the construction of a railroad dock extension. The resulting tsunami surged across the harbour, causing one death and more than US$20 million in damage to the wharf, boats, and infrastructure.

Tsunami waves become faster and shorter as the ocean depth increases (they move at a speed equal to the square root of the product of the acceleration of gravity and water depth). Moreover, tsunamis can travel transoceanic distances with little loss of energy (the rate of a wave's energy loss is inversely related to its wavelength). In the deep ocean, the 26 December 2004 Indian Ocean tsunami was barely noticeable, causing only a small and slow rising and falling of the sea. However, as the tsunami approached land, contact with the sea floor caused the waves to slow down and become more compressed, concentrating the wave energy in a smaller volume of water. Several giant waves with fast-moving, turbulent flows devastated the Indian Ocean coast: the rapid accelerations and decelerations associated with these fast-moving bores and surges are amongst the most destructive natural forces on Earth. The damage was exacerbated by strong currents and floating debris (large rocks, boats, etc.) that moved inland with great force, destroying homes and buildings, and drowning (or crushing) people in their path. The tsunami damage varied widely due to the dependence of wave velocity upon water depth and the complex submarine topography of Indian Ocean coastal regions. The extent of the Indian Ocean tsunami inundation zone depended primarily upon the existence (and characteristics) of objects such as buildings or vegetation which exert drag upon the base of the surges or bores. Where coastal zone bathymetry was optimal, the tsunami wave generates a breaking wall of water as high as 30 m (100 feet), as shown in Figure 3.

Human Impacts of the Indian Ocean Tsunami

Over 283 000 perished from the 26 December 2004 Great Sumatra–Andaman Earthquake, its subsequent tsunami, and resultant floods. The massive waves devastated hundreds of coastal Indian Ocean towns, stripping buildings and trees from coastlines. Children were particularly vulnerable to the surging waters and hence constitute one-third of the total deaths. Women were killed at a higher rate than men as they were often looking after children near the coast. In addition, many nations from outside the Indian Ocean region were also affected, as up to 9000 foreign (mostly European) tourists lost their lives. Germany and Sweden alone each reported more than 500 dead or missing. States of emergency were declared in Sri Lanka, Indonesia, and the Maldives, where about half of the homes were damaged. The Indonesian province of Aceh, on the northern tip of Sumatra island, was one of the most affected areas (Figure 3). Here, nearly 170 000 have been killed or are missing. Scientists investigating the damage in Sumatra documented wave heights of 20–30 m (65–100 feet) at the island's north-west end (Paulson, 2005).

Sri Lanka suffered the second greatest number of deaths, with over 30 000 confirmed killed and hundreds of thousands displaced from their homes, with the worst devastation occurring on the nation's south and east coasts. There were approximately 10 000 confirmed deaths in India, mainly in the southern state of Tamil Nadu, and thousands more missing, nearly all of them on the Andaman and Nicobar islands. The Thai government

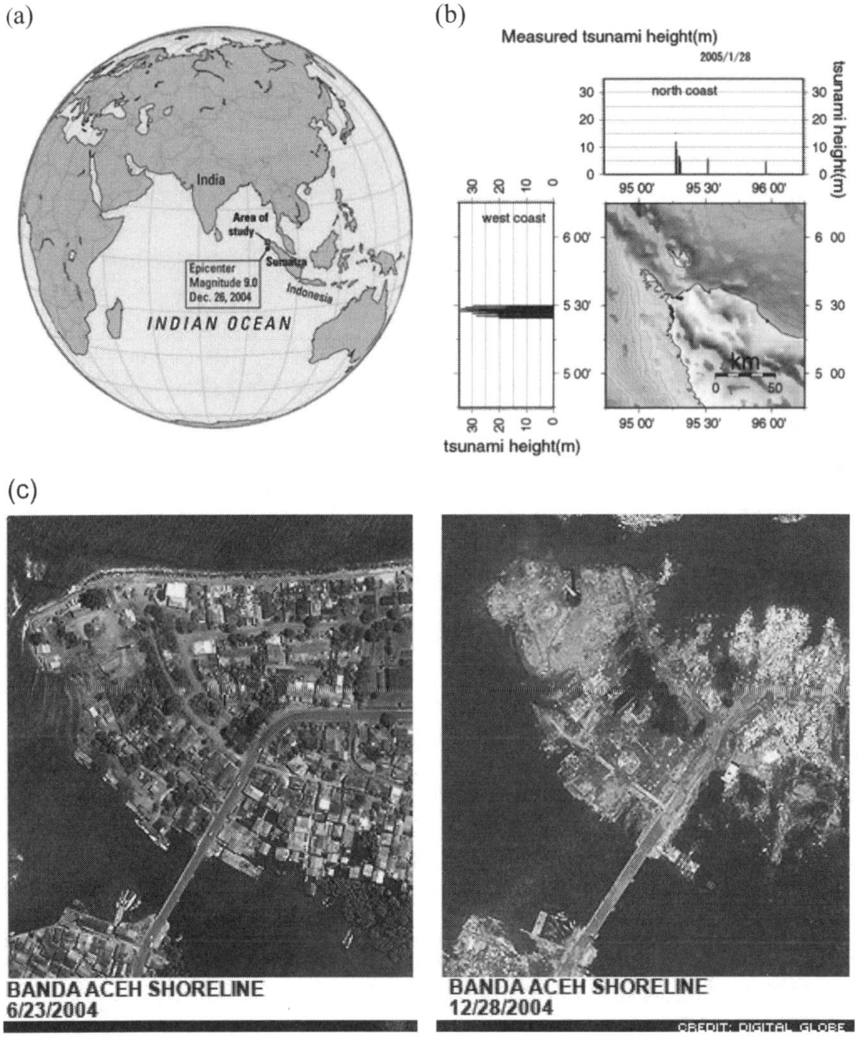

Figure 3. Damage in Banda Aceh, Indonesia. (a) Location of 26 December 2004 tsunami event. (b) Tsunami wave heights along the shoreline of Banda Aceh. (c) Tsunami damage: the image on the left shows the shoreline of Banda Aceh on 23 June 2004. The image on the right was taken on 28 December 2004, 2 days after the tsunami. *Source:* USGS (2005).

reports 5395 deaths and nearly 3000 missing, all in the six southern provinces facing the Andaman Sea. In Somalia, more than 4500 km (2800 miles) from the epicentre of the earthquake, entire coastal villages and communities were devastated. The island state of the Maldives suffered 82 deaths and 26 reported missing. The death toll in Myanmar is officially 61, but the figure may be higher (Figure 1). Coasts that have a land mass between them and the tsunami's location of origin were somewhat protected from the tsunami's impacts; however, tsunami waves did diffract around such land masses. Thus, the Indian state of Kerala was hit by the tsunami despite being on the western coast of India, and the western coast of Sri Lanka also suffered substantial impacts. Also distance alone did not ensure safety: Somalia was hit harder than Bangladesh despite being much farther away (the earthquake proceeded less quickly in the northern rupture zone, reducing the energy of water displacements near the Bangladeshi coast). Despite proximity to the earthquake epicentre, Malaysia suffered relatively fewer deaths than other Indian Ocean nations, many of which are thousands of miles further away.

In response to the tsunami tragedy, the United Nations (UN) has embarked on its costliest relief operation ever. Billions of dollars in aid assistance have been pledged by 90 donor nations, international agencies, and private donors over the next several years. This has already provided more than 1 million victims with food, emergency shelter, and medical assistance. Indonesia alone is expected to spend around US$5 billion in aid, in order to provide housing for hundreds of thousands of its citizens within the next 2 years. The UN World Food Programme provided food aid to more than 1.3 million people (more than 1 million are receiving food assistance in Sri Lanka and Thailand) and expects to feed about 800 000 people in Indonesia at least until the end of 2006. There were scattered reports of diarrhoea, malaria, dengue, measles, pneumonia, tetanus, and skin infections, but the massive humanitarian assistance prevented the large-scale outbreak of disease and starvation.

In the last 100 years alone approximately 800 tsunami events around the globe have been recorded. Many of these tsunami events have led to immense loss of life and the large-scale destruction of property. For example, the great Lisbon earthquake of 1755 generated a tsunami that destroyed large sections of Portugal and caused over 60 000 deaths. This is shown in Table 1, which documents the deadliest Atlantic, Mediterranean, and Pacific tsunami events in recorded history. Tsunamis are very common over the Pacific Ocean because it is surrounded on all sides by a seismically active belt. For example, the 1868 tsunami that occurred in northern Chile led to over 25 000 deaths. Devastating tsunamis have affected all the main Japanese islands, destroying hundreds of villages. In the 20th century alone, at least six major earthquake-generated tsunamis have occurred in Japan, including the 1896 event on the main island of Honshu (killing over 27 000) and the March 1933 tsunami in Sanriku (approximately 3000 deaths). Additional destructive tsunami events in Japan occurred in 1944, 1946, 1960, and 1983. The Hawaiian islands have also been repeatedly affected by tsunami events, such as the April 1946 event that inundated the city of Hilo and led to 171 deaths.

Table 2 provides a listing of all major tsunami events in the Indian Ocean since 1850. The first three columns in Table 2 provide data relevant to the date of the tsunami event. The fourth and fifth columns provide information about the Indian Ocean country, state, province, or island where the tsunami source occurred. The sixth column provides the earthquake magnitude (valid values range from 0.0 to 9.9) while the seventh column provides information on the focal earthquake depth (in km). The eighth column (deaths) provides information on the numbers of deaths resulting from the tsunami. This figure

Table 1. Deadliest tsunami events by region

Atlantic Ocean tsunamis			Mediterranean Sea tsunamis			Pacific Ocean tsunamis		
Date	Source	Deaths	Date	Source	Deaths	Date	Source	Deaths
1 November 1755	Portugal	60 000	1410 BCE	Greek islands	100 000 +	22 May 1782	Taiwan	40 000
7 June 1692	Jamaica	2 000	28 December 1908	Italy	10 000 +	20 September 1498	Japan	31 000
3 October 1780	Jamaica	300	6 February 1783	Italy	1 500 +	28 October 1707	Japan	30 000
7 May 1842	Haiti	300	11 January 1693	Italy	1 000 +	15 June 1896	Japan	27 122
6 December 1917	Nova Scotia	200	20 September 1867	Greece	12	13 August 1868	Chile	25 674
4 August 1946	Dominican Republic	100	16 October 1979	France	10	27 May 1293	Japan	23 024
7 September 1882	Panama	65	13 December 1990	Italy	6	21 May 1792	Japan	15 030
11 October 1918	Puerto Rico	42	9 July 1956	Greece	4	29 August 1741	Hokkaido	15 000
18 November 1929	Newfoundland	29	20 October 1859	Greece	2	24 April 1771	Ryukyu Islands	13 486
7 April 1934	Norway	17	11 September 1930	Italy	2	May 1765	China	10 000

Source: Tsunami Laboratory of Novosibirsk (2005) and USGS (2005).

may also include deaths from the earthquake that triggered the tsunami. Table 2 shows that Indonesia is particularly vulnerable to tsunamis (generated by earthquakes and volcanoes): sitting on a Ring of Fire, the islands in the archipelago lie so close to colliding tectonic plates that waves may hit land before information arrives from deep-water buoys. For example, the August 1883 eruption of the Krakatau (Krakatoa) volcano in the Sunda Straits (between Sumatra and Java) generated a massive tsunami wave that killed over 36 000 Indonesians, causing an entire island to virtually disappear.

The historical record of tsunami events dates back to antiquity, including the 79 AD eruption of Vesuvius (the oldest recorded tsunami in Italy). Following the 1755 Lisbon earthquake, systematic efforts to catalogue the locations of earthquakes (and the resulting tsunami events) were undertaken. Tsunamis leave their mark in historical, archaeological, and geological records in the form of features such as tsunami sand sheets in coastal marshes and turbidite deposits (a sand, silt, and mud sheet with highly distinctive characteristics). However, tsunami records face problems of completeness (significant areas such as Australasia were sparsely populated), distortion, and ambiguity, with respect to both the magnitudes of the tsunamis and their exact sources. However, recent advances in sea-bed mapping and the dating of tsunami deposits are enhancing the historical record.

Economic Impacts of the Indian Ocean Tsunami

Social and economic vulnerability to water hazards will probably increase significantly in the coming years as more than 75% of humanity is expected to live in coastal areas by 2030. Moreover, coastlines now contain infrastructure elements such as port facilities and power plants; damage to these critical infrastructures can cause large-scale social disruption and economic losses. The Indian Ocean tsunami severely affected tourism and fishing infrastructure (boats, fishing gear, etc.) in many countries, endangering the livelihoods of hundreds of thousands of families and leading to a loss of important foreign exchange (Kurien, 2005). Moreover, both the Sumatra–Andaman earthquake and the resulting tsunami may have significantly modified water depths in parts of the Straits of Malacca, one of the world's busiest shipping channels. Compiling new navigational charts and recharting altered ports and coastlines throughout the region will be costly and time-consuming.

The Indian Ocean tsunami impacted three major categories of critical infrastructure systems: utilities and lifelines; transportation networks; and communications. First, tsunamis had a catastrophic impact on power-generating stations and their support infrastructure (transmission networks and lines, water and sewerage systems, etc.). For example, in many parts of Sumatra, this led to interrupted electricity, petroleum, and gas supply systems. Destruction of treatment facilities, storage areas, pumping stations, and pipe networks in turn affected water and sewerage provision. Second, the Indian Ocean tsunami caused significant devastation to coastal transportation networks and facilities (railroad terminals, highways, airports, etc.). Roads covered with rocks and other debris typically became unusable while unpaved coastal road networks were often completely washed away. In addition, bridges, tunnels, airports, and viaducts located on low-lying coastal plains were destroyed, hampering relief work and the movement of goods and people. Facilities located directly on the coast such as harbour and port infrastructure (e.g. harbours, navigation aids, boats, fishing fleet stock) were among the first to be affected. Indirect losses were also significant, including the costs associated with the dredging of sediments deposited by tsunamis in harbours and navigation channels. Third, the Indian

Table 2. Historical tsunami events in the Indian Ocean region

Year	Month	Day	Country	Name	Magnitude	Focal depth	Deaths
1852	11	11	Indonesia	Sibolga, Sumatra	6.8		0
1861	2	16	Indonesia	South-west Sumatra	8.5		905
1861	3	9	Indonesia	South-west Sumatra	7.0		1 700
1861	4	26	Indonesia	South-west Sumatra	7.0		0
1861	9	25	Indonesia	South-west Sumatra	6.5		0
1863	3	16	Indonesia	Java, Indonesia			0
1868	8	19	India	Andaman Islands			0
1881	12	31	Indian Ocean	Bay of Bengal			0
1882	1		Sri Lanka	Sri Lanka			0
1883	8	26	Indonesia	Krakatau			0
1883	8	27	Indonesia	Krakatau			36 500
1884	2		Indonesia	Krakatau			0
1885	7	29	Indonesia	Ajerbangis	6.8		0
1885	12	14	Indonesia	Banda Aceh			0
1886	1	31	Indonesia	Koeta Radja			0
1886			Indian Ocean	Bay of Bengal			0
1889	8	16	Indonesia	Java–South Java			0
1896	10	10	Indonesia	South-west Sumatra	6.8		0
1907	1	4	Indonesia	South-west Sumatra	7.6		400
1908	2	6	Indonesia	South-west Sumatra	7.5	130	0
1909	6	3	Indonesia	Sumatra	7.7	40	0
1914	6	25	Indonesia	Indonesia	7.5		0
1917	1	21	Indonesia	Bali Sea	6.5	33	0
1921	9	11	Indonesia	South Java Sea	7.5		0
1922	7	8	Indonesia	Lhoknga, Aceh			0
1926	6	28	Indonesia	South-west Sumatra	6.7		0
1928	3	26	Indonesia	Krakatau			0
1930	3	17	Indonesia	Java–South Java			0
1930	6	19	Indonesia	Java–South Java Sea	6.0	33	0
1930	7	19	Indonesia	South Java Sea	6.5	33	0
1931	9	25	Indonesia	South-west Sumatra	7.5	3	0
1935	5	31	India	India	7.5		0
1935	12	28	Indonesia	South-west Sumatra	7.9	60	0
1936	8	23	Malaysia	Malay Peninsula	7.3		0
1941	6	26	India	Andaman Sea	8.1		5 000
1948	6	2	Malaysia	Malay Peninsula	6.5		0
1949	5	9	Malaysia	Malay Peninsula	6.7		0
1955	5	17	Malaysia	Malay Peninsula	7.2		0
1957	9	26	Indonesia	South Java Sea	5.5		0
1958	4	22	Indonesia	South-west Sumatra	6.5	33	0
1963	12	16	Indonesia	Java, Indonesia	6.5	64	0
1964	4	2	Indonesia	North-west coast	7.0	133	0
1967	4	12	Malaysia	Malay Peninsula	7.5	58	0
1981	12	31	Indian Ocean	Bay of Bengal			0
1982	2	24	Indonesia	Java Trench	5.4	52	0
1985	4	13	Indonesia	Bali Island	6.2	99	0
1994	2	15	Indonesia	South Sumatra	7.0	23	7
1994	6	2	Indonesia	Java, Indonesia	7.2	18	238
2000	6	18	India	Indian Ocean	7.8	10	0
2002	9	13	India	Andaman Islands	6.7	21	0
2004	12	26	Indonesia	Sumatra Coast	9.1	30	283 000

Source: USGS (2005).

Ocean tsunami caused the loss of communications systems (including radio and telephone network systems and satellite/fibre-optic communications) due to power losses and direct damage to buildings, cables, and transmission networks.

The indirect economic costs of tsunami events are often immense. For example, on the night of 17 July 1998 a 7.1 magnitude earthquake (with an epicentre near the coast of northern Papua New Guinea) triggered a series of waves up to 15 m high, devastating coastal villages along a 25 km stretch of coastline, and killing approximately 2000 people. Subsequently, the Papua New Guinea government closed the economically important Sissano lagoon (until April 1999) due to the potential health risks (the outbreak of disease, etc.). Some tsunami events have been followed by social unrest, giving rise to indirect military and police costs: in 1999, Turkish soldiers were needed to control looting and restore law and order in some regions after an earthquake-generated tsunami. On the other hand, after the 26 December 2004 Indian Ocean tsunami, citizens in Sri Lanka helped the government of Sri Lanka to reduce costs in the immediate aftermath of the disaster. Thousands of people, on their own initiative, engaged in direct individual humanitarian assistance, such as searching for tsunami survivors, transporting victims to hospitals, and delivering medical supplies, food, shelter, and clothing. Individual Sri Lankans and Sri Lankan volunteer organizations also reduced government expenses by collecting donations, organizing the purchase and delivery of relief goods, and undertaking reconstruction efforts (renovating schools, homes, etc.).

Environmental Impacts of the Indian Ocean Tsunami

Before the 26 December 2004 tsunami, many reef areas around the Indian Ocean were deliberately destroyed (they were considered impediments to shipping) and coastal mangrove trees were removed (to allow for more coastal residential developments). This deliberate destruction of coral reefs and mangroves caused coastal communities to become more vulnerable to the impacts of tsunami events: coral reefs and coastal mangrove forests protect people and their property by absorbing some of the force of the waves. For example, Surin Island's intact reefs off Thailand's coast protected people and property from the ravages of the Indian Ocean tsunami. However, the Sumatra–Andaman earthquake and the resulting tsunami significantly affected ecosystems over more than 1300 km (890 miles) of Indian Ocean, including mangroves, coral reefs, forests, coastal vegetation, plant biodiversity, and groundwater. Delicate Indian Ocean species (including coral, algae, and other marine invertebrates) and nearshore habitats were often devastated.

In some communities, salt water flooding contaminated drinking water supplies and farm fields for years to come. It is estimated that this may cause some Indian Ocean islands to become permanently uninhabitable, or dependent on outside food and water (Pearce, 2005). In the long term, the salt invasion may prove almost as destructive to the land as the tsunami itself. The tsunami may also have adversely affected the diversity of coastal ecosystems, already suffering from pollution and degradation. Together, salt water contamination, toxic chemicals, and sewage have polluted inland rice paddies and water supplies. Specifically, sand, sea, and earth entered drinking wells, salt invaded aquifers through porous rock, and a layer of salt was deposited over arable land. This situation is particularly acute in the Maldives, where over a dozen inundated coral reef atolls may remain uninhabitable for decades. Salted-over soil loses its fertility and leads to the death of plants and soil micro-organisms (in Sri Lanka alone, thousands of salt-affected rice, mango, and banana plantations may never recover).

Sedimentation is a significant impact of the Indian Ocean tsunami on the nearshore marine environment, including nearshore estuaries, mangrove and sea grass habitats, coral reef ecosystems, and sea turtle nesting areas. Sediments and excessive debris, including appliances, vehicles, fishing gear, boats, and buildings and other items, may bury entire ecosystems, affecting everything from corals and algae to fish and benthic substrates. Some of these items contained hazardous (non-biodegradable) chemicals that will affect the health of the shallow marine ecosystem for decades. Moreover, lines, nets, and other debris from damaged boats have the potential to entangle, trap, and drown protected species (marine mammals, turtles, etc.). It is known that degraded shallow water benthic habitats are less able to provide shelter and food for fish and benthic organisms.

Pacific Ocean Tsunami Warning Systems

A tsunami warning system detects tsunamis and issues warnings in order to foster tsunami-resilient communities. This requires a network of sensors (for tsunami detection) and a communications infrastructure (for issuing timely evacuation alarms). All tsunami warning systems feature multiple lines of communications (often using hardened dedicated systems) to alert population centres (with sirens, telephones, weather radios, emergency-alert systems, etc.) and to send emergency messages to disaster operations centres. The NOAA's Weather Service hosts the International Tsunami Information Center (ITIC) in Honolulu, Hawaii. The ITIC collects and disseminates tsunami information in order to improve tsunami preparedness. The current international warning system in the Pacific, the International Tsunami Warning System (ITWS), operates as a branch of the NOAA's National Weather Service. The ITWS is designed for the analysis of earthquake-generated tsunamis. It takes advantage of the fact that seismic (earthquake) waves travel at least an order of magnitude faster (within the Earth) than deep-water tsunami waves. Hence, the ITWS is capable of providing Pacific countries with 3–14 h warning before the onset of destructive tsunami waves. The primary monitoring activity involves tide gauge observations (sea-level changes) close to the earthquake source, as shown in Figure 4. When a tsunami is predicted, tsunami arrival times are estimated based on past experience and travel time models.

The NOAA's PTWC was established in Ewa Beach, Hawaii, in 1948 after the 1946 Aleutian Islands tsunami resulted in 165 casualties in Hawaii and Alaska. Today, the Richard H. Hagemeyer PTWC provides warnings for tsunamis to most Pacific Ocean nations, particularly the shorelines of North America, Hawaii, and Japan. The PTWC will issue advisory messages when major earthquakes occur in the Pacific Rim. The type of message (information bulletins, warnings, or watches) posted on the PTWC website depends on the interpretation of the seismic data received. A PTWC tsunami watch is undertaken when a 7.5 or greater magnitude earthquake is detected. The PTWC may convert a tsunami watch to a Pacific-wide tsunami warning bulletin after confirmation that a tsunami has the potential to cause damage to any populated coastal area within the Pacific basin at distances greater than 1000 km from the epicentre. Subsequent bulletins are issued at least hourly to update sea conditions (sea-level gauge readings, tsunami wave activity, etc.). A tsunami warning cancellation is issued when significant wave activity has subsided. Four NOAA agencies contribute to the US Tsunami Program: the NOAA's National Weather Service, NOAA Research, the National Ocean Service and the National Environmental Satellite, Data, and Information Service. Japan spends an estimated US$20 million a year on its extensive tsunami-alert structure: a network of fibre-optic sensors records seismic activity and passes the information to the Japanese Meteorological Agency (JMA), which rapidly estimates the height, speed, destination, and arrival time for earthquake-generated tsunamis (an alarm can be sounded within 2 min of an earthquake).

Historically, the tsunami operational decisions of the NOAA's PTWC and West Coast and Alaska Tsunami Warning Center depended primarily on seismic information and coastal tide gauge measurement (Gonzales *et al.*, 2005). However, the interpretation of these indirect data can be problematic (seismic/hydrodynamic coupling is poorly understood and shelf, coastline, and harbour features can lead to challenges in understanding tide gauge data). In addition, tsunamis may destroy tide gauges and, even if the gauge survives, the data are not reported until the tsunami reaches the coast. In 1995 engineering advances by the NOAA's Pacific Marine Environmental Laboratory enabled the creation of the NOAA tsunameter, a breakthrough for operational hazard assessment, tsunami research, and hazard mitigation.

In particular, the NOAA Deep-ocean Assessment and Reporting of Tsunamis (DART) system acquires and delivers direct tsunami measurements at deep-ocean locations from the tsunami source to the coastal communities. This is expected to significantly improve early detection and real-time measurements of deep-ocean tsunamis as they propagate toward population centres on the coast. The development and field testing of the first generation of tsunameters has allowed for the successful establishment of a Pacific tsunameter network (Milburn *et al.*, 1996). As shown in Figure 5, the NOAA DART tsunameter integrates four components into a single system: a bottom pressure recorder; an acoustic link; a surface buoy; and satellite telecommunications. Specifically, tsunameters record the pressure induced on the ocean bottom by passing tsunami waves and use acoustic telemetry to transmit these data to a nearby ocean surface buoy. Finally, the tsunami information is sent from the buoy to the warning centres via satellite.

Indian Ocean Tsunami Warning Systems

Despite a lag of up to several hours between the earthquake and the impact of the 26 December 2004 tsunami, many victims were taken by surprise as there were no tsunami warning systems in the Indian Ocean to detect tsunamis or to warn the general populace. A large majority of people in countries bordering the Indian Ocean were not taught to identify the tell-tale signs of the tsunami disaster (ground shaking, withdrawal of the sea from the coast, etc.). One of the few coastal communities to evacuate ahead of the Indian Ocean tsunami was located on the Indonesian island of Simeulue, extremely close to the epicentre of the Sumatra–Andaman earthquake. An earthquake and tsunami in 1907 have been incorporated into Simeulue Island folklore. Reminded daily by traditional folklore about tsunami dangers, the majority of the island's 70 000 people were able to flee to inland hills after the initial shaking, in advance of the actual tsunami (of course tsunamis can strike thousands of miles away, where the earthquake is felt only weakly or not at all).

Hence, in the aftermath of the Sumatra–Andaman earthquake and the subsequent tsunami disaster, Indian Ocean nations made efforts to improve their logistics, evacuation plans, and communications in order to provide a vital grid of communication and warning that was absent when the tsunami struck on 26 December 2004. Several Indian Ocean nations have already begun establishing national shoreline monitoring systems to mitigate the impact of future tsunami disasters. For example, in May 2005, the government of Thailand opened a US$2.5 million 24 h national disaster warning centre north of Bangkok, with links to the PTWC in Hawaii, the JMA, and the USGS. The centre is now able to warn coastal communities, such as Phuket Island, within a 3–5 min period, using radio, television, mobile phones, sirens, and towers. The government of Thailand expects to install sea-bed monitors and deep-water buoys around the Andaman Sea soon. In addition, Indonesia and Malaysia are using the powerful loudspeakers on village mosques to broadcast tsunami warnings while India is providing tsunami training for first responders and mobilizing police to evacuate threatened communities. However, challenges remain when delivering tsunami warnings to isolated regions where mobile phones and even televisions are scarce.

On 5 August 2005, the Indian Ocean tsunami intergovernmental co-ordination group for the Indian Ocean Tsunami Warning and Mitigation System (IOTWS) concluded that effective tsunami management requires a combination of three elements: an improved seismographic network; a real-time sea-level observing network (of ocean buoys and sensors) covering all of the Indian Ocean basin; and the deployment of deep-ocean pressure sensors capable of detecting the tsunami signal as it travels over the deep ocean. The backbone of this co-ordinated system will be an array of high-technology sensors measuring waves, tides, and water pressures. This involves the installation of a series of pressure gauges on the Indian Ocean sea floor to accurately and quickly detect the characteristics of an approaching tsunami. The IOTWS is expected to be fully operational (all national systems operating as a network) by July 2006, when at least 20 DART devices will be deployed. In addition, nearly 30 solar-powered sea-level tidal gauges, broadcasting data to satellites, will be installed by the end of 2005. Both the interim system and a final version will be comprised of a network of separate alert systems operated (and manned around the clock) by individual Indian Ocean nations.

Tsunami Education

Effective tsunami disaster planning and emergency-management preparations require appropriate tsunami education programmes: the public must be informed about the dangers and characteristics of the tsunami hazard and tsunami warning signs (the retraction of water from shorelines, shaking ground, loud rumbling, etc.). For example, the US state of Hawaii has undertaken a number of measures to increase tsunami awareness. First, tsunami signs inform

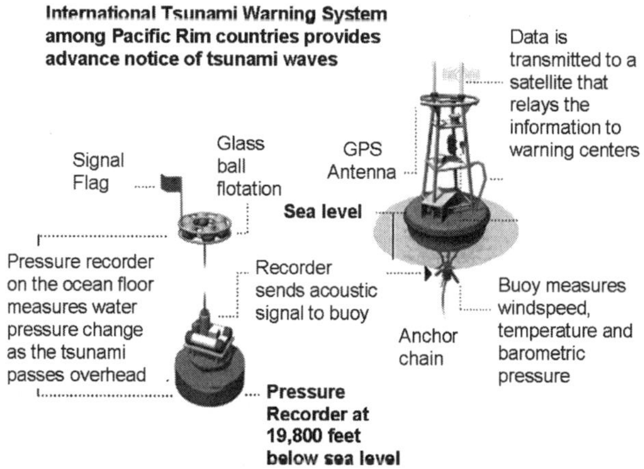

Figure 4. The ITWS. *Source:* NOAA (2005).

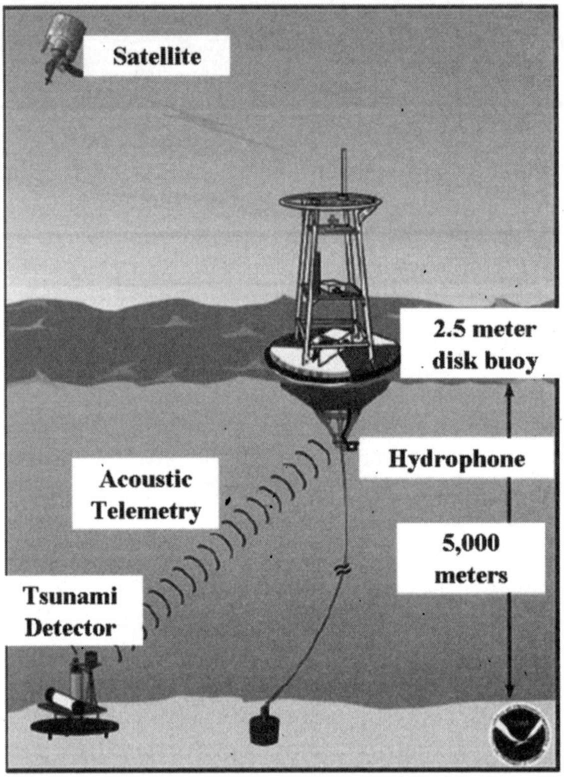

Figure 5. NOAA DART tsunameters measure tsunami waves in the deep ocean. *Source:* NOAA (2005).

beach goers about the dangers of tsunami waves and what action to take. Second, there is a system of state-wide emergency sirens along with tsunami drills held twice a year. Third, thousands of 18 min DVDs have been distributed warning surfers about the dangers of tsunami waves. In addition, the US National Tsunami Hazard Mitigation Program (NTHMP) seeks to promote 'tsunami-resilient communities' (Dengler, 1998) which should: (1) understand the tsunami risk; (2) have the knowledge and technology to mitigate the risk; (3) share tsunami hazard information; (4) exchange this information with other vulnerable communities; and (5) institutionalize integrated tsunami risk management (Jonientz-Trisler *et al.*, 2005). NTHMP mitigation products to promote tsunami-resilient communities include signage (tsunami hazard zone signs, evacuation signs, and educational signs) as shown in Figure 6.

Other mitigation material includes evacuation brochures, newsletters, and outreach products for a number of tsunami planning issues (surviving a tsunami, planning and designing for tsunami hazards, and warning systems procedures). Unfortunately, citizens of Indian Ocean nations were not aware of the tsunami warning signs: as the tsunami exposed large patches of the sea floor, many villagers and tourists collected stranded fish and took pictures, unaware that the tsunami would return with enormous speed and power. Tsunami education should emphasize that tsunamis are highly destructive (homes and small buildings located in low-lying coastal areas are not designed to withstand tsunami impacts), that they can occur at any time (day or night), and that they move rapidly inland (travelling up rivers and streams that lead to the ocean). Moreover, educational materials

Figure 6. Tsunami warnings.

should emphasize that precautionary measures differ according to whether the tsunami is detected on-shore, in the water, or on a boat. If a tsunami warning is heard while on land, it is important that people are evacuated to a safe facility outside the evacuation zone. The higher floors of multi-storey, strongly constructed buildings (reinforced concrete hotels, etc.) can be used if it is not possible to move quickly to higher ground further inland. On the other hand, if the Earth shakes while at sea on a boat, it is best not to return to port. This is because tsunamis are barely noticeable in the open ocean but can cause rapid changes in water level in ports and harbours (harbour officials should be contacted to determine when conditions are safe for navigation and berthing).

Summary and Conclusions

Prior to the 26 December 2004 Sumatra–Andaman earthquake and subsequent tsunami, Indian Ocean nations generally undervalued the importance of tsunami hazard planning and management due to the relative rarity of tsunami events in the region and the lack of resources. Accordingly, many nations were completely unprepared and the level of Indian Ocean tsunami readiness varied significantly. Moreover, tsunami simulation and modelling remain a complex challenge: tsunamis can be generated by a number of mechanisms (distant earthquakes and local seismic events, submarine landslides, and volcanic eruptions) and are influenced by many variables (including the amount of vertical and horizontal motion of the sea floor and the efficiency with which energy is transferred from the Earth's crust to the ocean water). Also, it is not easy to avoid dependency on external agencies for disaster recovery.

However, reducing casualties and property damage is possible by establishing more tsunami-resilient societies. This requires improved tsunami mitigation, preparation, response, and recovery. There are a number of timely and valuable steps that Indian Ocean governments can take to promote tsunami resilience. First, timely and convenient access to disaster information (educating stakeholders about tsunami hazards, etc.) and livelihood recovery assistance (health care, food assistance) is essential. Second, tsunami-resilient communities can be nurtured by improving ecological integrity (ecological systems such as coral reefs and mangrove forests can mitigate tsunami impacts), promoting institutions for disaster prevention (schools, training, etc,), and addressing poverty. For example, entire village communities in coastal areas were forced to live in poorly constructed dwellings (shacks, huts, etc.), often located directly on the coast, that offered little protection against the ocean waves. Third, tsunami resilience requires an appreciation of the cultural context: every Indian Ocean community has its own techniques for communicating with and caring for its citizens. Local tsunami mitigation strategies and recovery plans should consider self-identified community needs through meaningful consultation and participation. For example, how do local groups plan for disaster reconstruction and what are the traditional practices for dealing with loss? Fourth, tsunami resilience can be improved by harnessing recent technological innovations in tsunami detection, modelling, and warning. There are a number of promising scientific tools now available. For example, high-resolution digital bathymetric maps of continental shelf areas provide more accurate tsunami propagation paths, improving wave run-up estimates and arrival times at key locations. Fifth, Indian Ocean officials must integrate tsunami warnings into a multi-hazard system (which includes floods, storms, droughts, and other rare events), thereby increasing the preparedness of coastal communities, while reducing costs. Sixth, the proposed IOTWS requires an improved seismographic network, a real-time sea-level observing network covering the entire Indian Ocean basin, and the deployment of deep-ocean pressure sensors. Finally, in summary, if tsunami recovery is to be sustained indefinitely, effective community consultation and participation are required that empower stakeholders and stimulate meaningful dialogue.

References

Bilham, A. (2005) A flying start, then a slow slip, *Science*, 308, pp. 1126–1127.

Broad, W. J. (2005) Deadly and yet necessary, quakes renew the planet, *New York Times*. 11 January. Available at http://www.nytimes.com/, accessed August 2005.

Center of Excellence in Disaster Management (2005) Available at http://www.coe-dmha.org, accessed August 2005.

Cook-Anderson, G. & Beasley, D. (2005) NASA details earthquake effects on the Earth. Available at http://www.nasa.gov/home/hqnews/2005/jan/HQ_05011_earthquake.html (accessed 10 January).

Dengler, L. (1998) Strategic implementation for tsunami mitigation projects. NOAA Technical Report ERS PMEL-113 (PB99-115552). Available at http://www.pmel.noaa.gov/pubs/PDF/deng2030/deng2030.pdf, accessed August 2005.

Gonzales, F. I., Bernard, E. N., Meinig, C., Eble, M. C., Mofjeld, H. O. & Stalin, S. (2005) The NTHMP tsunameter network, *Natural Hazards*, 35, pp. 25–39.

Hanson, B. (2005) Learning from natural disasters, *Science*, 308, p. 1125.

Jonientz-Trisler, C., Simmons, R. S., Yanagai, B. S., Crawford, G. L., Darienzo, M., Eisner, R. K., Petty, E. & Priest, G. R. (2005) Planning for tsunami-resilient communities, *Natural Hazards*, 35, pp. 121–139.

Knight, K. (2005) Asian tsunami seabed pictured with sonar, New Scientist, 10 February. Available at http://www.newscientist.com/article.ns?id=dn6994, accessed August 2005.

Kurien, J. (2005) Indian ocean tsunamis devastate fisherfolk. Available at http://www.ukabc.org/tsunamis.htm, accessed August 2005.

McKee, M. (2005) Power of tsunami earthquake heavily underestimated, *New Scientist*, 9 February. Available at http://www.newscientist.com/article.ns?id=dn6991, accessed August 2005.

Milburn, H. B., Nakamura, A. I. & Gonzalez, F. I. (1996) Real-time tsunami reporting from the deep ocean, *Proceedings of the Oceans 96 MTS/IEEE Conference, Fort Lauderdale, FL*, pp. 390–394.

NOAA (2005) Available at http://www.noaa.gov, accessed August 2005.

Paulson, T. (2005) New findings super-size our tsunami threat: 80-foot waves blasted Indonesia. *Seattle Post-Intelligencer Reporter*, 7 February. Available at http://seattlepi.nwsource.com/local/211012_tsunamiscience07.html, accessed August 2005.

Pearce, F. (2005) Tsunami's salt water may leave islands uninhabitable, New Scientist, 5 January. Available at http://www.newscientist.com/article.ns?id=dn6840, accessed August 2005.

Pearce, F. & Holmes, B. (2005) Tsunami: the impact will last for decades, *New Scientist*, 15 January, p. 2482.

Press, F., Siever, R., Gretzinger, J. & Jordan, T. H. (2004) *Understanding Earth* (New York: Freeman).

Schechner, S. (2004) Earthquakes vs. the Earth's rotation: how major tremors alter the planet's wobble. Available at http://slate.msn.com/id/2111443/, accessed August 2005.

Schlesinger, W. H. (1997) *Biogeochemistry: An Analysis of Global Change* (New York: Academic Press).

Tsunami Laboratory of Novosibirsk (2005) Available at http://tsun.sscc.ru/tsulab/20041226fat.htm, accessed August 2005.

USGS (2004) Preliminary earthquake report, Available at http://earthquake.usgs.gov/recenteqsww/Quakes/usslav.htm, accessed August 2005.

USGS (2005) Available at http://neic.usgs.gov/neis/eqlists/eqsmosde.html, accessed August 2005.

Floods as Catalysts for Policy Change: Historical Lessons from England and Wales

CLARE L. JOHNSON, SYLVIA M. TUNSTALL & EDMUND C. PENNING-ROWSELL

Introduction

The tied relationship between flood disasters and the demand by the public for a policy response is well known (Tobin & Montz, 1997). In fact, it is arguably the case that it takes a severe and damaging flood to place flooding on the political agenda, at a time when the public and media response is such that a failure to act is politically unacceptable (Parker, 2000). There is, however, no guarantee that the nature of the policy issues raised by a major flood disaster will offer anything more than post-event response and recovery. For as Wilkins (2000, p. 84) observes in discussion of the 1993 and 1997 floods in the USA:

> If federal response to disaster relief is driven by the immediacy of an event, rather than by rational decision-making, the effort to put everything back to the way it was may increase future risk rather than relieving it and finding a long-term solution to risk reduction. In the wake of some disaster relief, and under the pressure of the media effect, the nation may have subsidized some poor decisions and penalized some sound proposals, forgoing opportunities for change.

Thus, state involvement not only has the capacity to implement options for risk and vulnerability reduction but may also both 'create' and 'promote' risk, particularly where such options are politically reactive as a result of a flood disaster. To reduce such reactivity, a greater understanding of the relationship between flood disasters and the policy-making process is required. At present, relatively little research has been conducted that seeks to explore why certain ideas dominate the flood policy agenda process, and are acted upon following a crisis, whilst others are not.

One reason for this is the limited number of longitudinal studies that seek to examine the impact of extreme events on policy evolution. Another is that as time elapses following a crisis, so too does the ability of the researcher to seek cause and effect correlations (Tobin & Montz, 1997). However, without exploring generalities across different contexts and crises, our ability to understand the process of policy change, as a result of these events, becomes diminished.

Drawing on the public policy theoretical literature, the paper begins by offering a theoretical framework for exploring flood policy evolution (Figure 1). By simplifying the policy-making process in this way, the complex reality of policy change has been categorized as either incremental or catalytic change. For the purposes of this paper, the analysis focuses on the relevance, or otherwise, of the catalytic nature of major flood disasters for changing policy. Thus, whilst the authors recognize that flood policy evolves as a function of incremental and catalytic drivers unconnected with major flood events, these issues are only examined in so much as they influence the context within which catalytic change occurs: not least important of which are the incremental changes in policy.

In England and Wales, the authors have identified three key phases of incremental change since World War II: 'land drainage', 'flood defence' and, most recently, 'flood risk management' (Tunstall *et al.*, 2004). Each of these phases reflects a changing set of beliefs, values and attitudes towards the flood problem, which in turn influences attitudes towards structural flood defences, flood warning systems, public awareness raising, land-use planning and development control for flood risk areas (Table 1).

Understanding this incremental change provides an important context for evaluating the policy impact of the four most significant floods in England and Wales in the post-war era: the 1947, 1953, 1998 and 2000 floods. The analysis of these floods is then used to qualify the authors' theoretical insights into floods as catalysts for policy change, which moves beyond the linear hazard–dose–response relationship so often associated with disasters.

Policy Change: A Theoretical Framework

Simplifying the processes leading to changes in public policy has long dominated the thinking of political scientists, from early writers such as Harold Lasswell (1930, 1948), Charles Lindblom (1959) and David Easton (1953, 1965) to more contemporary thinkers such as John Kingdon (1984, 2003) and Paul Sabatier (1987, 1991, 1993). Underlying these debates is the general agreement that policy changes as a consequence of changes in human behaviour. It is, however, precisely the debates concerning the causes of this human behaviour that differentiate the theoretical approaches to the study of policy, and indeed are why there could never be a single theory of policy change. Rather, the study of public policy requires an integrated approach, one which utilizes insights from a range of approaches when seeking to understand policy making, implementation and change (John, 1998).

To achieve such an integrated framework, the authors draw on insights from a range of theoretical positions, which include the work of Paul Sabatier and colleagues on policy advocacy coalitions (Sabatier, 1991, 1999; Sabatier & Jenkins-Smith, 1993), John Kingdon's policy streams approach (1984, 2003) and Baumgartner & Jones's (1993)

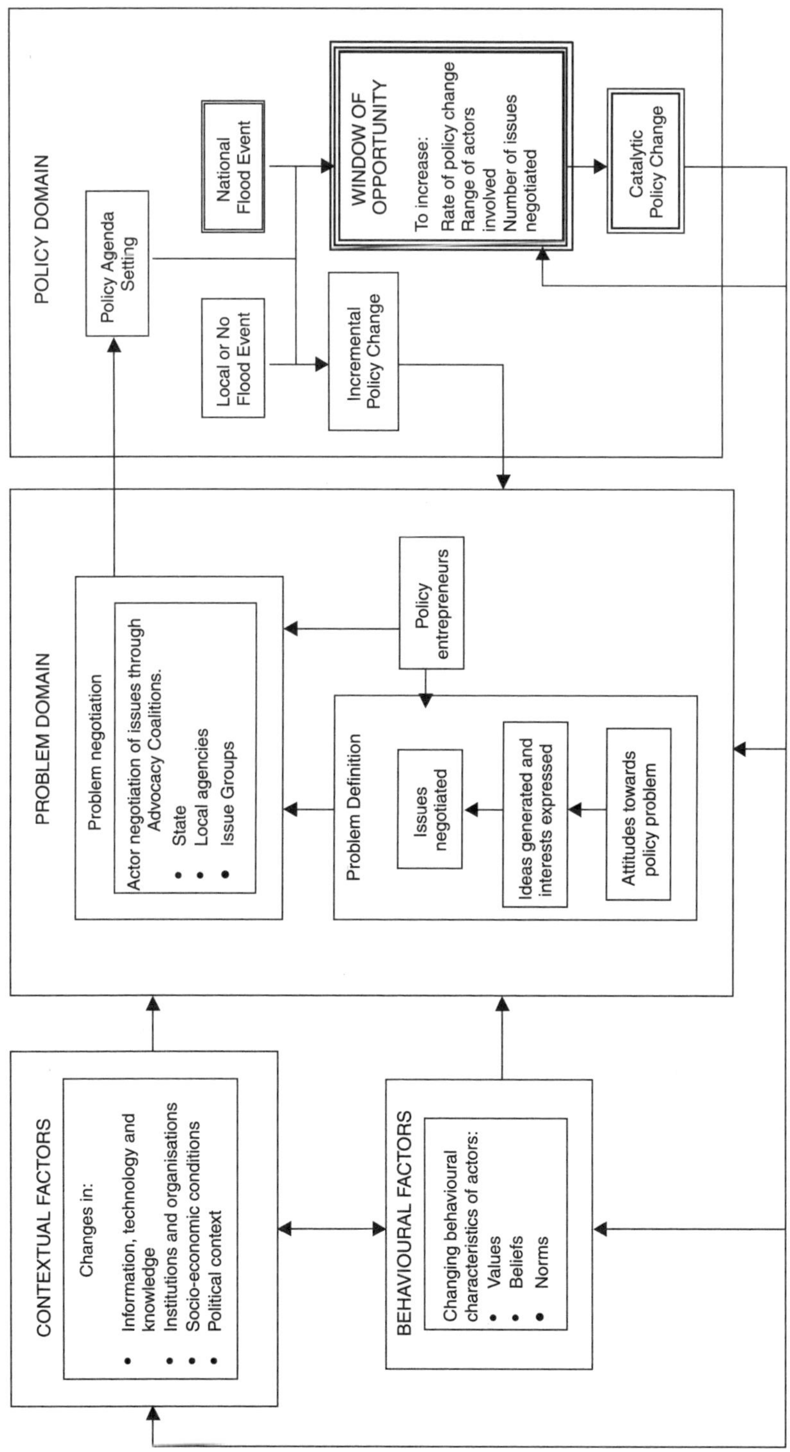

Figure 1. A framework of flood policy evolution.

Table 1. Key phases of incremental policy change, World War II to present day

Belief system	Land drainage (World War II–1970s)	Flood defence (1980s–1990s)	Flood risk management (2000s–)
Nature of humans	Humans have dominion over nature. Land is there for human use.	Humans have dominion over nature (power and right to exercise it).	Humans are part of nature, not superior to it. Nature has intrinsic value.
Priority of values	Priority on agricultural productivity and food security within the national economic context.	Priority on economic growth, national security and welfare standards.	Ecological and enironmental values should be viewed on a par with economic growth, national security and welfare standards.
Fundamental policy position	To improve and protect agricultural land from flooding.	To defend people and property from flooding.	To manage flood risks equitably and in accordance with the principles of ecologically sustainable development.
Basic policy mechanism.	Investment in land drainage and rural flood defence.	Investment in urban flood defence and flood alleviation schemes according to national priority criteria and economic appraisal processes.	A focus on decisions that satisfy social and economic needs whilst maintaining the ecosystem enhancement.

punctuated equilibrium approach. This has resulted in the development of a theoretical framework of incremental and catalytic policy change (Figure 1).

At its heart, the framework argues that the policy process is one of relative stability, 'punctuated' by times of rapid change (Baumgartner & Jones, 1993). Thus, the framework assumes incremental changes in policy when there is a local, or no, flood event and catalytic changes in policy when there is a national flood event (Lindblom, 1959; Baumgartner & Jones, 1993). The difference between the incremental and catalytic is then the rate of policy change, the extent to which new participants are attracted to an issue and the manner in which these new actors are mobilized by policy entrepreneurs (Baumgartner & Jones, 1993).

In the authors' framework, this punctuation is primarily brought about by flood events which, if national in nature, offer a 'window of opportunity' to increase the rate at which policy changes, the range of actors involved and the number of issues negotiated (Kingdon, 1984, 2003). From the policy streams approach, a 'policy window' emerges when a problem is recognized as a public matter in need of attention, and policy entrepreneurs can offer potential solutions to the problem at a time when there is political and institutional receptivity to such a solution. In the authors' framework, major flood disasters offer such a 'policy window', which, provided it can be 'seized' by policy entrepreneurs, offers the potential for changing policy in the agenda process. Policy windows do not stay open for long and the extent to which they result in any policy change is largely dependent on the ability of the policy entrepreneurs to get their issue onto, and keep it on, the policy agenda. Hence, the central importance of these key actors in the authors' model. They are, of course, important at times of incremental and catalytic change.

Incremental and catalytic policy change is modelled as a continuous process of learning and adaptation whereby both behavioural and contextual factors influence the manner in which the flood hazard problem is defined, but they do not determine change *per se*. Rather, the framework assumes that institutions, organizations, technology, socio-economic conditions and politics are important contextual factors within which ideas float in the policy 'primeval soup' (Kingdon, 1984). This places the authors' thinking into a particular subset of policy theorizing that deviates from those theorists who regard institutions as the dominant force of policy decision making: those who regard socio-economic conditions as dominant and those who regard the individual actor as dominant.

Based on the work of Sabatier and his colleagues, problem negotiation is modelled in Figure 1 as a direct function of advocacy coalitions (Sabatier, 1987; Sabatier & Jenkins-Smith, 1993). The hypothesis is that policy negotiation occurs between actors in competing coalitions who share similar beliefs and values, which then translate into core policy objectives and instruments. At times of incremental change, these coalitions are influenced by changing contextual factors and the values, beliefs and norms of the actors involved in the policy subsystem.

This process is expected to continue during times of catalytic change; the difference is that the national flood event provides an opportunity for increasing the number of issues negotiated, for increasing the range of actors involved, for providing the contextual environment to enable policy entrepreneurs to push for their ideas to form solutions and, ultimately, for increasing the rate at which policy changes (Kingdon, 1984, 2003; Baumgartner & Jones, 1993).

One reason this is assumed to occur is because national-scale flooding offers an opportunity for including those who were previously apathetic to, or disengaged from, the policy domain (Marsh & Rhodes, 1992; Baumgartner & Jones, 1993). These 'new' actors bring 'new' ideas about the problem definition and 'new' ideas about the issues or solutions required in the agenda-setting process. As each advocacy coalition is deemed to have core normative beliefs that fundamentally determine policy beliefs, and instruments for addressing the problem, these 'new' actors are expected to align with those coalitions sharing similar belief systems or, where no such coalition exists, a new coalition will be formed.

In the analysis of the case studies which follows, the validity of those assumptions directly relating to 'floods as catalysts' is tested. In particular, the assumption that major floods offer a 'windows of opportunity' is evaluated, as is the importance of key actors and the assumption that it is key contextual and behavioural factors that are important.

However, with only one advocacy coalition in the ascendancy during each of the floods it has not been possible to validate the competitive nature of coalition negotiation. Rather, the nature of these coalitions provides a useful understanding of how the underlying values, beliefs and norms of society, at the time of each of the floods, influence post-flood discourse and action (Tunstall *et al.*, 2004).

Four Floods: Four Catalysts

The context in which each of the floods occurred was an important factor for influencing the range of issues which were debated in the post-flood policy discourse. At the time of the earlier floods, river management was focused on rural land drainage and flood defence to increase agricultural productivity, promote self-sufficiency and protect farm profitability. Structural and 'hard engineering' schemes dominated with little regard for environmental impact. At the time of the later floods, a more strategic, catchment-wide, multi-method and integrated approach to land and water management was characteristic. There was less emphasis on structured flood defences and more on flood warnings, public awareness, self-help, land-use planning and development control. Sustainable development was now a guiding principle, which meant that social, economic and, particularly, environmental issues had to be fully addressed in river, coastal and land management.

In each case study, a number of issues were raised, and debated, in the post-flood discourse. These included, amongst others, agriculture and land-use management, development control, flood warnings and public awareness raising, structural and non-structural defences, the suitability of institutional structures, funding arrangements, insurance, flood risk mapping, the lack of information, the need for research and the competence of relevant authorities. However, in each of the case studies, of the many issues that were raised and debated, only one or two were then translated into recognizable policy changes in the agenda-setting process:

- in 1947, policies that sought improvements in the structured defence of rural agriculture were accelerated;
- in 1953, it was the adoption of the 1953 standard for coast protection and the development of a storm tide warning system for the east coast;
- in 1998, policies that sought to improve and modify the system of flood warnings and public awareness raising were accelerated and implemented;
- in 2000, it was development control that dominated the policy agenda, resulting in the strengthening of the Planning Policy Guidance (PPG) 25 governmental guidance, already under negotiation prior to the 2000 floods.

The 1947 Riverine Floods

The cause of the 1947 flood can be attributed to three main factors: snow, ice and rain. Heavy snowfalls were followed by a deep depression bringing with it rain and increased temperatures. This led to rapid snow-melt, which, in combination with the frozen ground, resulted in the flooding of 690 000 acres (1 acre = 0.41 ha) of agricultural land and thousands of homes, businesses and industries. This flood has long been seen as a benchmark, unique in its extent, volume, persistence and impact; probably the worst flood the country had ever experienced and certainly the worst for many centuries (Barker, 1948).

Of the many issues that were raised and debated, it was policies which sought to improve the structured defence of agricultural land that were 'catalysed' by the 1947 flood (Johnson *et al.*, 2004). This was due to the combined influence of the flood characteristics, the context within which it occurred and the actors involved.

The flood characteristics were important, not because of the river and tidal floods which inundated 30 of the 40 counties in England and naturally receded over time, but because of the prolonged and widespread Fenland flooding in East Anglia. In this prime agricultural region, the floods were a combination of bank breaching and overtopping and the widespread saturation and flooding of the lowland fens themselves. This caused flooding which not only required human intervention through artificial drainage and pumping but also extended beyond the capacity of the systems to cope, leading to the prolonged and widespread agricultural damage for which the 1947 flood is remembered (Barker, 1948).

This damage occurred at a time of rapid political, social and economic change. The most significant of these were the emergence of the nation from World War II, the global economic recession, world food shortages and the dollar crisis. The national economy was under significant economic stress, was dependent on domestic agricultural production for food consumption and was facing a serious fuel and power crisis. The war context meant that the catchment boards, whose responsibility it was under the Land Drainage Act 1930 to implement flood prevention measures, had been prevented from doing so due to shortages in labour, material and financial resources. This context was a key factor in the dominance of agricultural issues, and structured flood prevention, in the post-flood discourse.

This discourse was also influenced by the values, beliefs and norms of the actors involved, which, at the time of the flood, were dominated by the land drainage coalition (Table 1). The prioritization of land drainage and flood protection meant that, although politically powerful, the farming lobby was not a key player in the post-flood discourse *per se*. Rather, their cause was taken up by those within government who neither debated nor raised any questions about the suitability of the agricultural policy response that emerged. Instead, it was the government, through the Ministry of Agriculture and Fisheries (MAF), in which responsibility for managing the policy response resided. This effectively determined which of the issues raised by the 1947 flood would be acted upon, not least because MAF did not have any direct powers to change institutional structures or impose flood warning obligations or development control restrictions. It did, though, have the power to influence agricultural policy and provide incentives to the catchment boards for structural works.

This it enacted, not through a public inquiry, but by convening conferences with the various catchment boards based on reports by the chief engineers of these boards. Formally, these conferences were expected to provide the necessary information for determining long-term policy. Informally, the government firmly believed that it had enough information from its own knowledge for developing long-term policies and strategies. Therefore, in actuality, the government used the conferences to ensure that its goal to accelerate the long-term improvements in river defences would be met by the catchment boards, whose responsibility it was to draw up and implement such plans (Johnson *et al.*, 2004). Here, therefore, we see the importance of the mechanism by which the flood is evaluated and the formal powers of the actors involved.

The 1953 East Coast Floods

The 1953 flood was a unique flood event in recent history in its nature, severity and impact. Its fundamental cause was meteorological with an exceptional and widespread northerly gale, which developed in the North Sea and spread southwards down the eastern coast of Scotland and England towards the Dutch coast. This gale resulted in significant rises in sea levels coinciding with spring tides. The resultant storm surge caused overtopping and breaching throughout the eastern seaboard from the Humber in the north to the Thames estuary in the south on 31 January and 1 February 1953. Whilst not the greatest surge on record, this was, and remains, the greatest flood in recent history. The impact of the flood was immense by UK standards. Along the 1370 (2205 km) miles of defences, there were 1200 breaches, more than 300 people lost their lives, 160 000 acres of farmland were inundated, 24 000 houses were flooded and 32 000 people were evacuated from their homes.

Of the many issues that were raised and debated, it was the adoption of the 1953 standard for coastal protection and the development of a storm tide warning system for the east coast that were 'catalysed' by the 1953 flood (Johnson *et al.*, 2004). Here again we can see the combined importance of the flood characteristics, the actors involved and the context.

As in 1947, the war and post-war context were important factors in determining policy changes. For coastal protection, the lack of labour, finance and materials for the construction and maintenance of sea defences, plus the fact that many of the secondary defences had been removed, or destroyed, by military operations, meant that a focus on flood defence construction, standards and maintenance was inevitable. As with the 1947 case study, there was widespread consensus and agreement that this was an appropriate policy response.

Similarly, the context and nature of the flood were important factors in determining the centrality of science and warnings in the policy discourse. The most outstanding feature of the east coast floods was the total lack of warnings, particularly considering there was ample time available for such warnings. This was as much to do with the administrative structures of forecasting as it was to do with the total lack of any appropriate medium for warning dissemination. It was the Meteorological Office's responsibility to report and forecast on matters influencing the wind and weather patterns, and it was the Admiralty's Hydrographic Department's responsibility to report, and forecast, on issues concerning the sea and its tides. At the time of the flood, the latter department was issuing long-term forecasting data of little relevance to the developing surge and the former was supplying forecasts without any knowledge of sea conditions.

In addition, the policy outcomes were significantly influenced by the actors involved in the policy discourse. At the outset, responsibility for ensuring that such a national disaster would not occur again was 'placed' in the hands of the government and scientific community. The Home Office set up an internal departmental inquiry, chaired by Lord Waverley, supported by 13 'experts' and 'elites' from government and the scientific community. By acting as gatekeeper in this way, the Home Office set the committee's terms of reference and ensured that discussion would focus on science and understanding without focusing on recriminations. Indeed, it is interesting to note that the 'voice of opposition', in the guise of the local authorities, was not represented on the committee at all.

This committee was the primary vehicle for policy formation (Waverley, 1954). The most significant actors in this process were the scientists. The value of 'science' for transforming policy, with very little public scrutiny, was never questioned. The flood had illustrated that the country simply did not know enough about the mechanisms and magnitude of the flood hazard threat, let alone have the mechanisms for issuing warnings. Hence, by highlighting the inadequacies in information, knowledge and technology the flood had created a universal recognition for the need for further research. The scientists on the Waverley Committee were careful to show that they understood the event, whilst illustrating the need for improved knowledge about how to predict and protect. It is no coincidence that the two policy changes from the 1953 flood are exactly in these areas: prediction in the storm tide warning system and the standard of defence for coastal protection. The question was not whether to provide a warning system or coastal protection, it was how.

This led, firstly, to the development of an emergency warning system between April and September of the same year, a significant investment in research, the greater involvement of the Meteorological Office and, ultimately, the development of the storm tide warning system. These were not policy issues regarded as important before the flood.

Secondly, the committee's recommendation that flood defences should be raised to the level of the 1953 flood, in areas of high property value and agricultural importance, became the predetermined standard of protection for coastal defence structures, where economically justifiable. It is worth noting, however, that whilst the need to restore coastal defences to some standard was accepted, the use of the 1953 level as the standard did not emerge due to any scientific deliberation. The provision of a lower standard was debated in the deliberations of the Waverley Committee but it was politically impossible for them to suggest a lesser standard.

Finally, the 1953 flood led to particular attention being focused on the flood defence structures operating in the London estuary. Given the importance of this region, and the potential risk to which it was exposed, there was deemed to be too narrow a margin of protection. The result was the implementation of the Thames Barrier some 30 years later. This serves to illustrate how a flood as devastating as that experienced in 1953 can 'catalyse' policy developments many years into the future. In this case, it took 30 years to develop the necessary knowledge, information and technology to justify the economic expenditure required.

The 1998 Easter Flood

The widespread riverine flooding of Easter 1998 was due to a slow-moving, prolonged and extremely intense rainfall front which affected locations in a 5000 km^2 band across the middle of England and into Wales (Bye & Horner, 1998). At the time, the flooding was described as the "worst flood of the century" (Guardian, 1998, p. 8) and in many places was the most severe ever recorded (Bye & Horner, 1998). The floods were exceptional in their speed of onset, their night-time nature and their dispersal. They accounted, directly or indirectly, for the loss of life of five individuals and, on the basis of best estimates, caused widespread damage to 4500 properties, 522 businesses, 2000 caravans and numerous cars. In total, this flood caused approximately £300 million of damage (1998 prices).

As a case study, the 1998 flood is particularly useful for illustrating the significance of key policy 'champions' in the post-flood discourse. It is also useful for illustrating how the institutional and political context can be important factors.

At the time of the 1998 flood, the more affluent and environmentally conscious public, and to some degree the authorities, had become complacent about flooding. In the period since 1953, large sums had been invested in structural flood protection so that many of the locations at greatest risk were defended to some standard. At the same time, the populations living in coastal and riverine floodplains, both defended and undefended, had expanded so that by 1998 an estimated 5 million were at risk from flooding. Added to this, the fact that the flood occurred after a long period in which no major flood event had taken place meant that it took both the authorities and the public by surprise. The authorities had no experience in dealing with a large-scale flood event, and the public had become accustomed to a flood defence orthodoxy, with a belief system centred on the ability of humans to 'control' nature, to defend people and property from flooding and, ultimately, to avoid floods (Table 1).

The lack of experience of the relatively new institutions of governance for dealing with major flood disasters is of particular note, the newly elected Labour government having been in power for less than 1 year and the Environment Agency (EA) having been in existence for only 2 years.

The new government was important because, upon election, it began the process of 'shaking up' departments and policy making, including the division of the Ministry of Agriculture, Fisheries and Food (MAFF) that dealt with flood defence, driven by a fundamental desire to adopt a different approach to its predecessor. As a result, at the time of the flood, the data and justifications for policy and expenditure on flood defence were already under review by the House of Commons Select Committee on Agriculture's Special Session on Flood and Coastal Defence. Whilst this did not lead to any radical thinking for changing the institutional and funding arrangements, it did provide an important forum for discussion of the 1998 flood and its implications.

In the 2 years the EA had been operational, its preoccupation with establishing structures and procedures for dealing with its new responsibilities for waste management and integrated pollution control meant that relatively little attention had been paid to flood warning dissemination. Furthermore, the EA had only assumed lead responsibility, from the police, for issuing warnings directly to those at risk in September 1996. Thus, at the time of the flood, the EA inherited warning arrangements remained fragmented and ineffective.

Given this institutional and governance context, it is unsurprising that the performance of the EA, particularly in its failure to issue warnings, was severely criticized by the public and Members of Parliament in the post-flood discourse. In response, and partly to forestall demands for a formal and potentially more politically damaging public inquiry, the EA set up an independent review of its performance. This was the key mechanism for evaluating potential policy changes following the 1998 flood. It was led by two respected individuals, Michael Horner, the director of an engineering consultancy, and Peter Bye, a former local authority chief executive who brought a fresh social perspective to the inquiry. Of particular note was the fact that the review team, free to decide how to conduct its inquiry, chose, as a priority, to visit the flooded locations and meet with flood victims. This personal commitment to flood victims and social issues, on the part of Peter Bye, was very different to the formal processes, and reliance on expert and scientific evidence, that had gone before.

The flood highlighted the need to plan for, and manage, all flood events including extreme events, which may become more common in the UK with climate change. It also demonstrated that it would not be possible to deal with all potential flood risks through structural flood defences. Thus, it gave added impetus to the transition towards a risk management approach to flooding (Table 1). The independent review team's report, the Bye Report (Bye & Horner, 1998), emphasized the human impacts of flooding and the need to give greater attention to the human and social aspects of warning systems and messages. This, it can be argued, had a significant and broad effect on the balance between technical and social issues in research, staffing, policy and its implementation within the EA, with greater emphasis given to social issues.

Ultimately, this flood led directly to changes in both flood warnings and public awareness raising. The Bye Report called for consistent national standards in the flood warning system and for an extension of some form of warning arrangement to cover areas not at high risk. It led directly to the development of a new, national and clearly defined four level system of flood warning codes with associated behavioural advice. A telephone message service, FloodLine, providing regularly updated information on flooding and flood warnings, was established. The codes and the FloodLine number are now used by the broadcast media in their weather reports. The Bye Report also contributed to the initiation, in 1999, of annual autumn flood warning campaigns, using national media advertising, as well as local information campaigns. These have, generally, had the effect, over the past 5 years, of raising and maintaining awareness of flooding and flood risk, of the EA as the body for flood defence, of the flood warning system and of the appropriate response to floods.

A further outcome of the 1998 flood was a review of the automatic voice messaging system that the EA had seen in 1996 as its main means of delivering warnings directly to the public. The flood demonstrated that the capacity of this locally based telephone message system would be inadequate to cope with a major disaster. Therefore, the EA initiated a project to develop and implement a multi-media warning dissemination service that would have the capacity to deliver a much larger number of warning messages using a wider range of communication channels including the telephone, internet, email and mobile phone and text messages. Reliance upon technological solutions for warning dissemination also raises questions as to the access to, and acceptability of, new technologies particularly for disadvantaged groups and those with disabilities. These issues are being explored in EA research.

The Autumn 2000 Floods

In the flood event of 1998, a key contextual factor was that there had not been a major national flood event for decades. Conversely, in 2000, it was the fact that there had been such an event only 2 years earlier that was significant. Indeed, the events of 1998 and 2000 had a cumulative effect on policy.

The more widespread, and in some locations repetitive, flooding of autumn 2000, which affected 827 locations in all regions in England and Wales, and a total of approximately 11 000 properties (Penning-Rowsell *et al.*, 2002), was the result of rainfall 'waves' unprecedented during the previous 270 years. However, following so closely on the event of Easter 1998, it could not be represented as 'exceptional'. It thus served to reinforce the view that it was essential to plan for extreme events which might occur as a result of climate change and that it was not possible to deal with all flood risks through structural flood defences.

The close occurrence of two major events was interpreted by some Members of Parliament as evidence of the impact of climate change in bringing more extreme weather events, although a later review of the evidence to support this was inconclusive (Department for the Environment, Food and Rural Affairs, 2001). The Deputy Prime Minister characterized the event as a 'wake-up call' to the UK. The much-improved performance in 2000 of the newly enhanced flood warning system reassured the government that this issue was being addressed. Thus, interest shifted to focus on another 'non-structural' policy approach: development control. The issue of new buildings in floodplains, themselves at risk from flooding, or giving rise to risks elsewhere, had been raised after the 1998 flood, with the government agreeing to review its existing advice to local planning authorities on development and flood risk. It had had no plans to do so prior to the 1998 flood.

After the autumn 2000 floods, the government decided to look again at the planning guidance it was in the process of drafting, in order to strengthen and toughen the message it contained. It indicated, in consultation, that it was looking for a 'step change' in the responsiveness of the land-use planning system to the issue of flood risk management.

The final PPG on *Development and Flood Risk* that was issued in July 2001 (Department for Transport Local Government and The Regions (DTLR), 2001), unlike the policy changes that emerged following the flood events of 1953 and 1998, was the outcome of a formal government process of policy review and revision. In this, the key actors were the government department with responsibility for planning (the Department for the Environment, Transport and Rural Affairs (DETR), and subsequently, the DTLR), the EA acting as a key adviser and MAFF acting in an advisory capacity. However, it can be argued that over the period following the two floods, the policy guidance was subjected to unprecedented scrutiny in Parliament by House of Commons select committees. It was also given exceptional exposure to formal public examination because there were, unusually, two consultation periods on draft versions before the issue of the final guidance. These consultations attracted substantial numbers of written responses, most of which were supportive of a strengthening of the guidance.

The final guidance, PPG 25, was a direct outcome of this elaborate process, introducing a different and much more rigorous risk-based approach towards development and flood risk. It embodied the precautionary principle and required an appropriate flood risk assessment for all development plans and proposals, where there was any flood risk on site or that might be generated elsewhere. It proposed a sequential test that should be applied, giving priority to those areas that presented little or no risk over those with low to medium or higher risk. It also differentiated between developed and undeveloped, defended and undefended high risk areas, and indicated that the standard of protection and life of flood defences and the type of development proposed should be taken into account in decisions on development. It also stated that development in "functional flood plains should be wholly exceptional" (DTLR, 2001). Thus, the guidance reflected the recognition of the importance of the land-use planning system to the flood risk management approach.

The Impact of Flood Disasters on Policy Change

The impact of major floods on policy change appears to be dependent on a combination of contextual, behavioural and environmental drivers. Factors of particular significance for determining the typology of possible policy responses have been found to be a combination of:

- the magnitude of the flood disaster, and its impact (environmental drivers);
- the availability of technology, knowledge and information at the time of the flood (contextual drivers);
- the socio-economic, political and governance structures in place (contextual drivers);
- the dominant attitudes, beliefs and values of society towards the flood hazard (behavioural drivers).

These, in combination, appear to be the critical factors for influencing the policy options available at the time of major flood disasters. Which of these options are then 'catalysed' into actual policy changes then appears to be influenced by the nature of the flood and its impacts, the mechanism employed in the policy evaluation process and the key actors involved.

In each of the case studies, only one or two policy options were translated into any catalytic change in policy towards flooding. Thus, the authors' analysis finds value in the theoretical arguments of Baumgartner & Jones (1993) that institutions and organizations can only focus on one issue at a time in the agenda-setting process. We are also in agreement with Baumgartner & Jones (1993), in that major floods do appear to 'punctuate' the policy process by offering periods of public interest, media scrutiny and policy action such that the rate of policy change is increased. However, in none of the case studies did these floods destroy dominant sets of images or institutions. Neither did the 'new' participants to the policy subsystems offer any destabilizing influence. Perhaps, therefore, major flood disasters offer a rather more toned down version of 'punctuation' than Baumgartner & Jones's (1993) theory would have us believe.

In all but one example (the 1953 storm tide warning system), the policy changes that were 'catalysed' by the flood did not reflect any 'new' policy ideas, or fundamental changes in policy direction. Rather, the floods served as a catalyst for increasing the rate at which a policy idea, already under consideration prior to the crisis, was given prominence and acted upon. In 1947, the flood merely served to reinforce the already firmly held belief in agricultural flood defence structures, in 1998 the flood served to highlight the failures of the EA to fulfil the flood warning duties that were already assigned to it, and in 2000 the flood served to strengthen the planning guidance which was already under review. Only in 1953 did any new policy idea emerge, brought about by the flood and its impact, which clearly illustrated the policy need.

In this sense, the authors' analysis appears to corroborate the theoretical ideas of Kingdon (1984, 2003), that major flood disasters offer a 'policy window' for change: a problem is recognized, and policy communities and entrepreneurs can press for their ideas to form solutions to the problem at a time when there is political receptivity to the defined problem and proposed solutions (Kingdon, 1984, 2003).

In each of the case studies, different mechanisms were favoured for evaluating the causes, impacts and lessons learnt from the flood, the choice of which was influential in determining the nature of the post-flood policy discourse, the issues raised and the key actors involved.

- In 1947, lessons were drawn, and policies formulated, primarily on the basis of the conferences between the chief engineers of MAF and the catchment boards. This ensured the prominent role of the Minister for Agriculture and key individuals in MAF and the catchment boards.
- In 1953, it was a departmental committee investigation, overseen by an independent chair (Lord Waverley), which deliberated on policy issues. This gave prominence to the Secretary of State for the Home Office and the scientists on the Waverley Committee.
- In 1998, the primary mechanism was the independent review inquiry commissioned by the EA to directly evaluate its own performance. It is no coincidence that key individuals in the policy process during this flood were the independent review team (Peter Bye and Michael Horner), and key individuals in the EA and MAFF, not least of which was the Minister for Agriculture himself (Elliot Morley).
- In 2000, it was the PPG 25 review, and consultation process, implemented by the DETR. This process ensured that key actors in this flood emerged from the DETR, MAFF and the EA as well as members of the Select Committee on Agriculture and the Select Committee on the Environment, Transport and the Regions.

Of the most prominent actors involved, the majority were already involved in the flood hazard policy domain. Certain actors were, however, by their own admission, 'new' to the flooding field, most notable of whom were Lord Waverley in 1953 and Peter Bye in 1998. However, in none of the case studies did the ideas of either the '*in situ*' or 'new' actors represented a radical shift in policy thinking. Rather, they were able to develop their policy ideas within a receptive environment. In this way, the case studies illustrate the importance of key actors in influencing which ideas influence policy change, but this influence can only be achieved where there is general consensus about the legitimacy of these ideas in the first place.

Thus, whilst we are in agreement with Kingdon (1984, 2003) that a national flood offers a 'policy window' for policy solutions to be attached to policy problems, the extent to which policy entrepreneurs lie in wait to 'seize the opportunity' is less clear-cut. It is arguably the case that in 1947, the chief engineer of MAF did so when pursuing the catchment boards for action, as indeed the EA did in 1998, and more particularly in 2000, in its exploitation of the opportunity to press for a strengthening of guidance on development and flood risk. However, the same could not be said of Peter Bye in 1998 or the scientists on the Waverley Committee in 1953. Thus, from the authors' case studies, it is difficult to assess whether it is policy entrepreneurs which 'soften up' policy communities or *vice versa*. As a result, the authors would argue for key actors to be characterized as policy 'champions' rather than policy 'entrepreneurs' (Kingdon, 1984, 2003). It is these 'champions' who put in the necessary resources to ensure a policy option gets onto, and stays on, the agenda. This requires a receptive environment.

The dominant beliefs, values and attitudes of this receptive environment are usefully categorized using the policy advocacy framework originating from the work of Paul Sabatier and colleagues (Sabatier, 1991, 1999; Sabatier & Jenkins-Smith, 1993). This usage is, however, not as the framework's authors originally intended: as a means of understanding inter-coalition bargaining. Rather, the present authors found value in their theory as a means of understanding the relationship between beliefs, values and policy positions. This provided a basic understanding of the dominant belief systems, policy positions and mechanisms at the time of each of the floods. In none of the case studies did the floods alter these core beliefs. Quite the contrary: it was the beliefs, values and norms at the time of the floods which impacted on the nature of the policy response which arose. The authors do, however, recognize that a more detailed analysis of policy change over many decades would be required for a thorough investigation of the value of advocacy coalitions to flood policy evolution.

Conclusions

The authors' case studies illustrate that major flood disasters do indeed provide a 'window of opportunity' for key actors, or policy 'champions', to accelerate the rate at which policy is implemented. It is, however, not the norm for any 'new' ideas to materialize. Rather, the policy ideas, which are accelerated during such events, appear to be dependent on the contexts that have gone before. In this way, the typology of possible policy responses changes over time and it is this change which provides us with potential 'signals' of policy responses in the future. Thus, our ability to predict future changes in policy as a result of major flood disasters is largely dependent on our knowledge of the issues, actors and ideas seen as important before the flood. By monitoring these incremental changes we should be able to anticipate, rather than react to, catalytic changes that may occur in the future. Key factors in this process appear to be a combination of environmental, behavioural and contextual drivers.

Acknowledgements

The authors would like to thank the Economic and Social Research Council (ESRC) for funding this research (award no. RES-221-25-0037) as part of the Environment and Human Behaviour New Opportunities Programme.

References

Barker, D. (1948) *Harvest Home: The Official Story of the Great Floods of 1947 and Their Sequel* (London: HMSO).

Baumgartner, F. R. & Jones, B. D. (1993) *Agendas and Instability in American Politics* (Chicago, IL: University of Chicago Press).

Bye, P. & Horner, M. (1998) *Easter 1998 Floods: Final Assessment by the Independent Review Team*, vol. 1 (Bristol: EA).

Department for the Environment, Food and Rural Affairs (2001) To what degree can the October/November 2000 flood events be attributed to climate change? Department for the Environment. Food and Rural Affairs final report FD2304, London.

Department for Transport, Local Government and The Regions (DTLR) (2001) Planning Policy Guidance note 25: development and flood risk. Available at http://www.odpm.gov.uk, accessed August 2001.

Easton, D. (1953) *The Political System* (New York: Alfred A. Knopf).

Easton, D. (1965) *A Framework for Political Analysis* (Englewood Cliffs, NJ: Prentice-Hall).

Guardian (1998) The flood of the century, 13 April, p. 8.

John, P. (1998) *Analysing Public Policy* (London: Continuum).

Johnson, C., Tunstall, S. & Penning-Rowsell, E. (2004) Crises as catalysts for adaptation: human response to major floods. Final report, ESRC Environment and Human Behaviour New Opportunities Programme, Flood Hazard Research Centre Publication 511. Available at http://www.fhrc.mdx.ac.uk/resources/publications.html.

Kingdon, J. (1984) *Agendas, Alternatives and Public Policies*, pp. 13–39 (Boston, MA: Little, Brown).

Kingdon, J. (2003) *Agendas, Alternatives and Public Policies*, 2nd edn (New York: Longman).

Lasswell, H. D. (1930) *Psychopathology and Politics* (Chicago, IL: University of Chicago Press).

Lasswell, H. D. (1948) *The Analysis of Political Behaviour: An Empirical Approach* (London: Kegan Paul).

Lindblom, C. E. (1959) The science of muddling through, *Public Administration Review*, 19, pp. 78–88.

Marsh, D. & Rhodes, R. A. W. (Eds) (1992) *Policy Networks in British Government* (Oxford: Clarendon).

Parker, D. J. (Ed.) (2000) *Floods* (London: Routledge).

Penning-Rowsell, E., Chatterton, J., Wilson, T. & Potter, E. (2002) *Autumn 2000 Floods in England and Wales: Assessment of National Economic and Financial Losses* (Enfield: Flood Hazard Research Centre, Middlesex University).

Sabatier, P. A. (1987) Knowledge, policy-orientated learning and policy change, *Knowledge: Creation, Diffusion, Utilization*, 8, pp. 649–692.

Sabatier, P. A. (1991) Toward better theories of the policy process, *Political Science and Politics*, 24, pp. 147–156

Sabatier, P. A. (1993) Policy change over a decade or more, in: P. A. Sabatier & H. Jenkins-Smith (Eds) *Policy Change and Learning: An Advocacy Coalition Approach*, pp. 13–39 (Boulder, CO: Westview Press).

Sabatier, P. A. (Ed.) (1999) *Theories of the Policy Process* (Boulder, CO: Westview Press).

Sabatier, P. A. & Jenkins-Smith, H. (Eds) (1993) *Policy Change and Learning: An Advocacy Coalition Approach* (Boulder, CO: Westview Press).

Tobin, G. A. & Montz, B. E. (1997) *Natural Hazards: Explanation and Integration* (New York: Guildford Press).

Tunstall, S. M., Johnson, C. L. & Penning-Rowsell, E. C. (2004) Flood hazard management in England and Wales: from land drainage to flood risk management. Paper presented at World Congress on Natural Disaster Mitigation, *19–22* February, *Conference Proceedings*, 2, pp. 447–454 (New Delhi: Vigyan Bhawan). Available at http://www.fhrc.mdx.ac.uk/resources/publications.html.

Waverley, J. A. (1954) *Report of the Departmental Committee on Coastal Flooding* (London: HMSO).

Wilkins, L. (2000) Searching for symbolic mitigation: media coverage of two floods, in: D. J. Parker (Ed.) *Floods*, pp. 80–87 (London: Routledge).

Quantified Analysis of the Probability of Flooding in the Thames Estuary under Imaginable Worst-case Sea Level Rise Scenarios

R. J. DAWSON, J. W. HALL, P. D. BATES & R. J. NICHOLLS

Introduction

A significant proportion of the world's population reside in the coastal zone: 1.2 billion people lived in the near-coastal zone (the area within 100 km distance and 100 m elevation of the coastline) in 1990 at densities about 3 times the global mean (Small & Nicholls, 2003). Urbanization is an important trend, including a high concentration of the world's biggest cities (Nicholls, 1995), and a considerable portion of global gross domestic product (GDP) is also produced in coastal zones (Turner *et al.*, 1996). Coastal development is already threatened by a range of hazards. Moreover, human development and activity, such as the destruction of saltmarsh, the construction of artificial barriers and development, are impacting the natural behaviour of the coastal zone and changing the risk of flooding and storm damage. Climate change, in particular sea level rise (SLR), is an additional pressure that will greatly impact on developments in the coastal zone (Nicholls, 2002).

There is a great deal of uncertainty as to future changes in sea level. The rate of SLR is governed by many factors (in some cases opposing). The heavily dampened response of the ocean to atmospheric warming means society is committed to a certain amount of SLR regardless of any other factors (Wigley, 1995; Nicholls & Lowe, 2004). Thermal expansion and ice cap melting will result in increased SLR, whereas Antarctica's growth and increased terrestrial storage of water resources by humans could act to reduce SLR. Regional oceanographic and meteorological effects and vertical land movements further compound the uncertainty about future relative SLR at any site. For London, the UK Climate Impacts Programme (UKCIP) has downscaled the global mean SLR scenarios of the Intergovernmental Panel on Climate Change (IPCC) (2001), suggesting a rise of between 0.26 m and 0.86 m, with an additional +50% uncertainty due to regional oceanographic effects (Hulme *et al.*, 2002). If Antarctica were a positive contribution to SLR during the 21st century, the rise in sea level could be larger than the above scenarios. In the extreme, the total collapse of the Western Antarctic Ice Sheet (WAIS) would result in an SLR of 5–6 m (Mercer, 1978; Oppenheimer, 1998). The probability and associated time-scale for such a collapse are highly uncertain. While significant collapse is considered highly unlikely to occur during the 21st century (Vaughan & Sponge, 2002), our limited scientific understanding of collapse does not allow us to discount total collapse over 100 years, although such an event must have a low probability. It is noteworthy that since this research project finished recent observations by the British Antarctic Survey have cast further doubts on the stability of the WAIS (Tirpak *et al.*, 2005). These extreme WAIS collapse scenarios were explored in the Atlantis project, which was primarily a study of response to abrupt climate change (Tol et al., forthcoming), and the present study supported this research with an analysis of possible impacts and potential responses for the Thames estuary.

Abrupt climate change is defined by the National Research Council (2002) thus:

> ...from the point of view of societal and ecological impacts and adaptations, abrupt climate change can be viewed as a significant change in climate relative to the accustomed or background climate experienced by the economic or ecological system being subjected to the change, having sufficient impacts to make adaptation difficult.

Hulme (2003) suggests that SLR in excess of the maximum IPCC (2001) estimates of approximately 1 m per century may be considered as an example of abrupt climate change. Previous analyses of the impacts of climate change events have often focused on broad-brush estimates of loss of GDP for different global policy scenarios such as reducing CO_2 emissions (e.g. Keller *et al.*, 2000; Mastrandrea & Schneider, 2001) or global impacts on populations across sectors (e.g. Arnell *et al.*, 2002; Parry, 2004). Analysis at a more local level enables more specific quantification of the impacts and consideration of possible responses. This paper presents an exploratory analysis of SLR scenarios in the range of 1–5 m over the next 100 years. The analysis is based on a two-dimensional hydrodynamic storm surge model of the Thames estuary. The results of a limited number of plausible adaptation scenarios are also explored.

London, the capital of the UK, is home to 7.5 million people and has an average population density of 4500 people/km^2. It is estimated that currently 1 million people (London Assembly, 2002) and 300 000 properties are in the (present day) tidal flood risk area. The dominant flood threat comes from surge tides, caused by areas of low pressure travelling south or south-west over the North Sea which funnel a bulge of water into the confines of the southern North Sea and hence the Thames estuary. London is defended by a complex system involving over 200 km of embankments and floodwalls, the Thames Barrier and a suite of warning systems (Environment Agency, 2003a, b). While the possibility of a barrier was discussed earlier in the 20th century, the decision to build the present defences was made in direct response to the 1953 storm surge, which killed 300 people and flooded 65 000 ha of low-lying land on the UK's east coast (Gilbert & Horner, 1984). A full 30 years (only 8 years of which were construction time) after this event, the barrier was completed. Since the construction of the Thames Barrier, London's previously derelict docklands have been regenerated with new homes and businesses, including the new financial district around Canary Wharf. Significant future development is planned in flood-prone zones alongside the tidal Thames in London, Kent and Essex over the next 15–30 years (Office of the Deputy Prime Minister (ODPM), 2004).

The design life of the Thames Barrier and associated defence system is until 2030, by which time it is expected that rising sea levels will reduce the standard of protection to below a 1:1000 year standard (i.e. the barrier is expected to be overtopped by the 1 in 1000 year event if it were to occur). Given the long lead-time required to upgrade the defences, planning of the flood risk management strategy until 2100 is already in its early stages. A number of possible management strategies in the Thames estuary are being considered by the Environment Agency (2003a, b, c)and the ODPM (2004). These consider landward realignment of the flood defence line as a complementary strategy to raising defences and the generation of wetlands, thereby reversing a long-term trend of encroachment and land claim into the tidal Thames. It is likely that the current defence system will be raised by ~1 m at a cost of roughly £4 billion (London Assembly, 2002), although much more analysis is required before detailed options and their costings become available.

Inundation Modelling of the Thames Tidal Floodplain

Tidal circulation, and hence inundation resulting from a storm surge, is driven by a number of physical processes including gravitational forcing, density variations due to salinity, turbulence and surface wind stress. Numerical models of such flows range in complexity from fully three-dimensional solutions of some derivative of the Navier–Stokes equations (Cugier & Le Hir, 2002) to models that treat flow as one-dimensional in the down-estuary direction only (Kashefipour *et al.*, 2002). Choice of model depends on the morphology of the estuary and the hydrodynamics of the flow to be simulated. The tidal Thames is a drowned river valley with morphology typical of coastal plain estuaries (Dyer, 1973) with extensive tidal mudflats. Simulation of inundation over low-gradient tidal floodplains with significant flood defence structures (embankments, etc.) requires at least a two-dimensional modelling approach with relatively high spatial resolution (grid scales 250 m or less) to represent the complex geometry. However, full two-dimensional modelling of the whole Thames estuary remains computationally prohibitive at this scale, particularly if one wishes to simulate multiple scenarios associated with different potential futures.

To reduce the computational burden of the hydrodynamic calculations for this study it was assumed that for very large storm surge events the gravitational forcing at the estuary mouth is the dominant driving process. The authors also assume that the large volume of saline water entering the estuary is such that density can be assumed to be constant. Lastly, the authors assume the flood wave propagation can be represented as an approximation to a two-dimensional diffusive wave. Here the estuary and floodplain are discretized as a grid of rectangular cells. Flow between cells is then calculated simply (Figure 1) as a function of the free surface height difference across each cell face:

$$Q = \frac{h^{5/3}}{n}\left(\frac{h^{i-1,j} - h^{i,j}}{\Delta x}\right)^{1/2} \Delta y \tag{1}$$

Change in water depth in a cell over time t is then calculated by summing the fluxes over the four cell faces:

$$\frac{\mathrm{d}h^{i,j}}{\mathrm{d}t} = \frac{Q_x^{i-1,j} - Q_x^{i,j} + Q_y^{i,j-1} - Q_y^{i,j}}{\Delta x \Delta y} \tag{2}$$

where $h^{i,j}$ is the water free surface height in cell (i,j), Δx and Δy are the cell dimensions, n is a friction coefficient and Q_x and Q_y describe the volumetric flow rates between floodplain cells. Equations (1) and (2) give similar results to a more accurate finite difference discretization of the diffusive wave Equation (Horritt & Bates, 2001) but with much reduced computational cost. This model, LISFLOOD-FP, has been shown to perform as well as full two-dimensional codes (Bates & De Roo, 2000; Horritt & Bates, 2001) for the case of fluvial flooding, whilst Dawson *et al.*, (2003) and Bates *et al.*, (2005) have achieved good performance for coastal flood modelling. This research has suggested that model resolution and topographic data quality are stronger controls on the ability to simulate flood inundation than model physics. The authors hypothesize that this representation of flooding as a volume-filling problem is also true for estuaries, and that a good first order model therefore only requires the simple routeing of the correct volume of

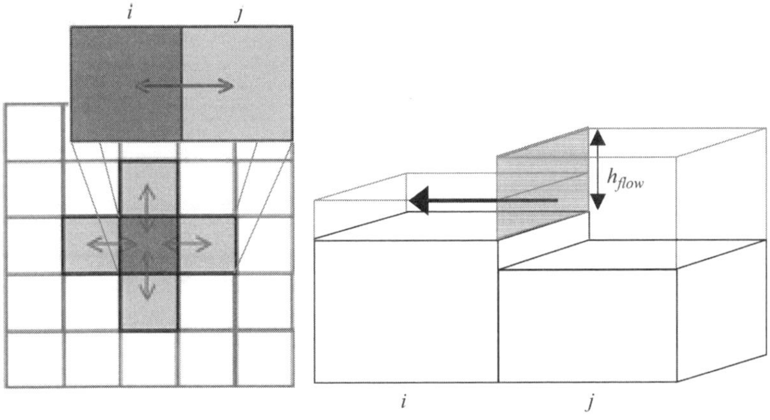

Figure 1. Representation of flow between raster cells in LISFLOOD-FP.

Table 1. Predicted water levels measured in metres above Ordnance Datum for given return periods along the Thames in 2030; sites are represented by triangles in Figure 3

	Return period (years)								
Site	2	5	10	20	50	100	200	500	1000
Silvertown (Thames Barrier)	5.21	5.44	5.61	5.77	5.95	6.11	6.29	6.51	6.70
Erith	5.10	5.33	5.50	5.66	5.87	6.02	6.21	6.43	6.62
Tilbury	4.66	4.89	5.03	5.19	5.37	5.53	5.70	5.91	6.09
Sheerness	3.89	4.08	4.20	4.33	4.49	4.62	4.77	4.96	5.12

Source: Jones (2001) and Environment Agency (2003a).

water over a detailed representation of the estuarine topography and flood defences. Effects such as wind shear, turbulence and density currents are, for very large storm surges at least, assumed to be secondary. While this simple model may not simulate fine details of the wave propagation (e.g. the timing of flood onset at a particular point), it will capture the maximum flood extent sufficiently well to evaluate the impact of different future SLR and defence scenarios.

Model boundary conditions are derived from statistical analysis of water levels in the estuary, shown in Table 1, and a typical storm surge history as shown in Figure 2. Therefore it is assumed that the storm surge characteristics captured in Table 1 and Figure 2 are constant over time, which, given the large magnitude of the SLR scenarios being considered, is a reasonable first assumption. The digital elevation map has been constructed from interferometric synthetic aperture radar data (Colemand & Mercer, 2002) and has a root mean square error of ± 0.7 m.

Flood risk in estuaries is dominated by two mechanisms: defence overflow and defence breaching (Hall *et al.*, 2003a). Overflow volumes of flood defences can be calculated using standard weir Equations (e.g. see Chadwick & Morfett, 1993). Consideration of defence breach scenarios adds a further computational burden on the assessment of inundation probabilities (Hall *et al.*, 2003a) and requires the (often controversial) assessment of failure probabilities and breach widths to flood defence structures. The assessment of the

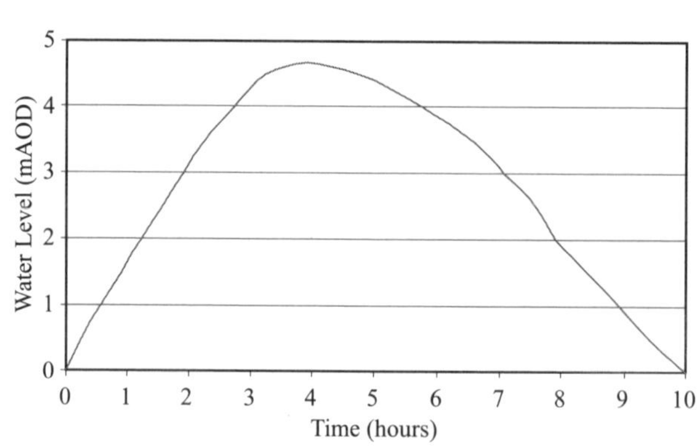

Figure 2. The 1953 storm surge as measured at Sheerness. *Source:* Rossiter (1954) and Smith & Ward (1998).

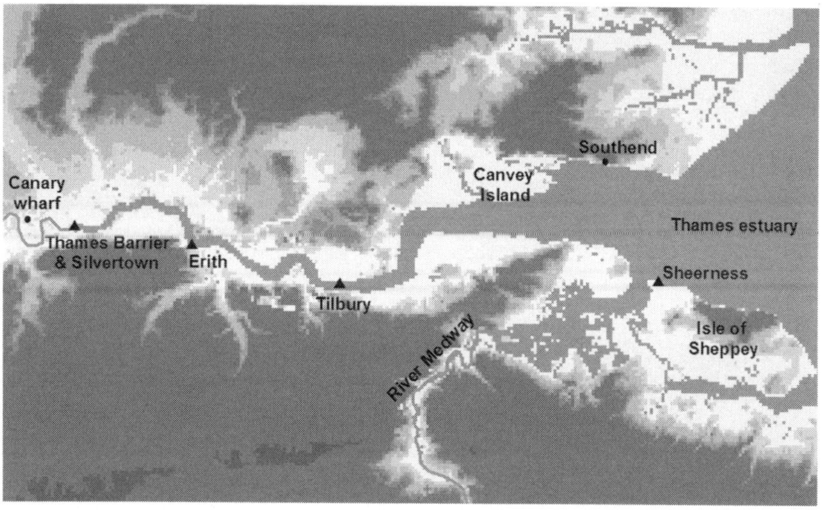

Figure 3. Map of Thames estuary showing the floodplain relief (lighter shades mean lower ground), the four tidal gauges (triangles), the Isle of Sheppey, Canvey Island and the River Medway.

flood risk contribution from defence breaching can be efficiently assessed using methods outlined by Dawson (2003). However, the good present condition of the barrier and defences downstream of it (Environment Agency, 2003a) (and the likelihood of significant upgrade in the next few decades) means that failure is unlikely except under extreme conditions of defence overflow. Due to the extreme nature of the SLR scenarios being considered in the modelling, the additional contribution to the total inundation volume from breaching is negligible when compared to the inundation volume from extreme overflow events.

A number of potential adaptation scenarios are considered in the hydrodynamic model. The first, a *status quo* scenario, gives a baseline for comparison against other possible intervention scenarios. This represents a policy of maintaining the strength of the defences at the 2003 crest level. This scenario is a baseline and it should be noted that significant upgrades to the current defence system are planned in line with the high-end UKCIP scenarios for SLR (~1 m crest level rise) (Hulme *et al.*, 2002). Another adaptation scenario is to raise the current defence system: in this case the defences are assumed to maintain the same relative levels of protection. The construction of new alternative downstream tidal barriers is also considered: the first is upstream of the Medway estuary at Canvey Island (E: 581250, shown in Figure 4); and the second is downstream of the Medway at the Isle of Sheppey (E: 593250, shown in Figure 5), a location first suggested by Gilbert & Horner (1984), who contributed substantially to the design and construction of the current Thames Barrier. When implementing the possibility of constructing an outer barrier, it was assumed that the other estuary defences would be upgraded to reflect the increased standard of protection offered by the new barrier.

Key results from the hydrodynamic model are presented. Figures 6–8 show the flood outline for the 1:1000 year flood event, assuming that no defences are constructed or raised, for 1 m, 3 m and 5 m net SLR scenarios. The topography of the Thames estuary floodplain is such that the area at risk from flooding increases most rapidly for the first 1 m

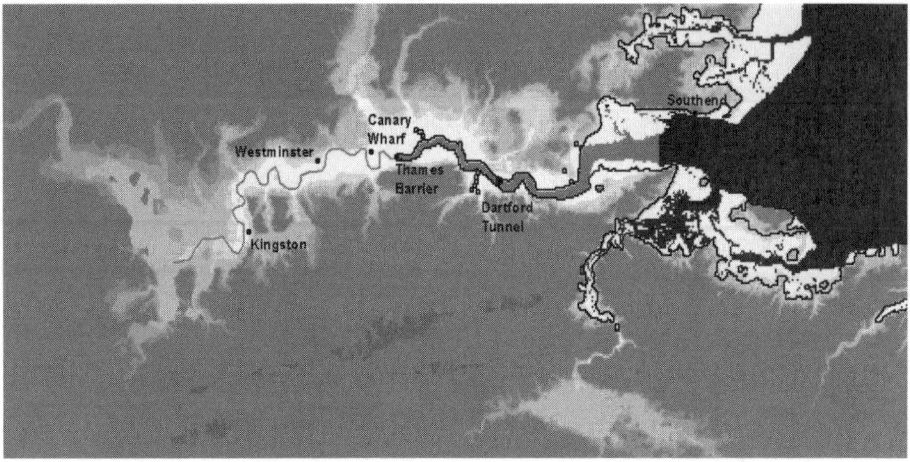

Figure 4. The 1:1000 year flood after 1 m net SLR assuming the Canvey Island Barrier is designed to a standard of protection of 1:1000 in 2030.

of SLR. Each additional 1 m of SLR produces a smaller increase in the inundated area. In the worst-case scenario of 5 m net SLR (equivalent to a 5 m SLR and no intervention, or a 6 m SLR and 1 m raise in crest defence level), approximately 1000 km^2 of land and 1 million properties would be at risk from flooding during the 1000 year event (Figure 8). The approximate area inundated for the 1 m and 3 m SLR scenarios is 650 km^2 and 850 km^2, respectively, whilst the current area at risk from the 1000 year event is $\sim$400 km^2. The number of properties likely to be inundated for different return periods and SLR scenarios is summarized in Figure 9. The most dramatic result is that even with a 2 m net SLR, 200 000 properties (assuming no further development) would be flooded annually assuming no adaptation response.

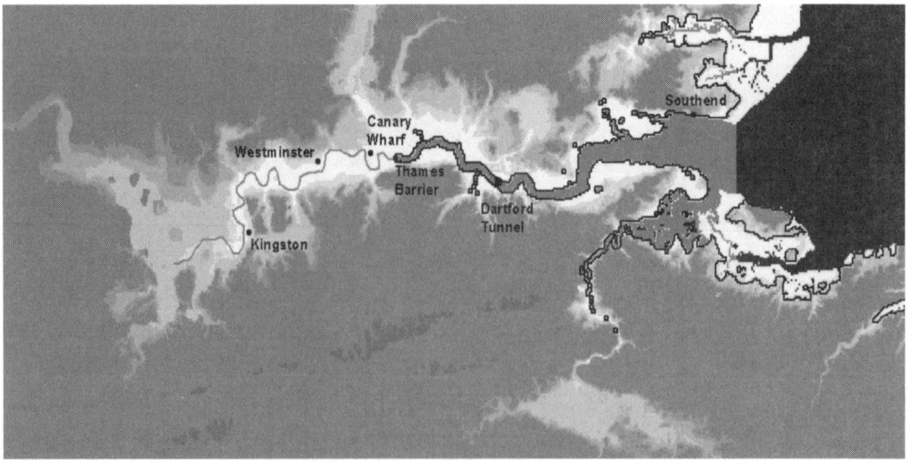

Figure 5. The 1:1000 year flood after 1 m net SLR assuming the Sheppey Barrier is designed to a standard of protection of 1:1000 in 2030.

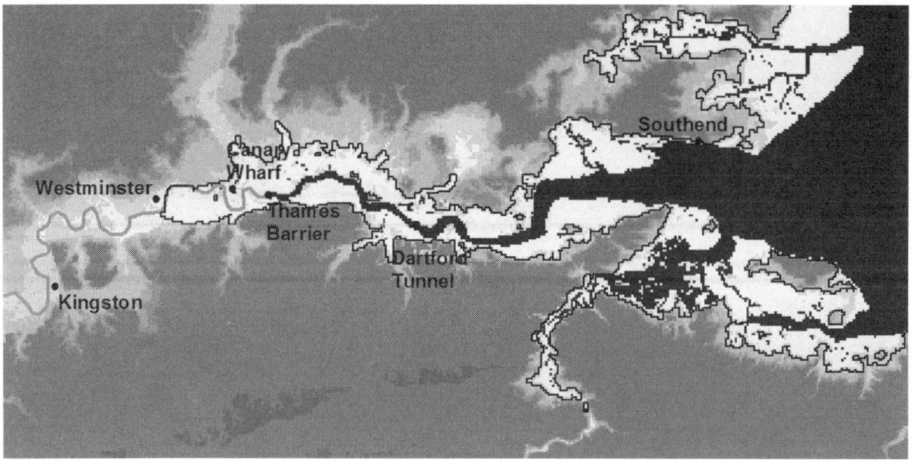

Figure 6. The 1:1000 year flood outline after 1 m net SLR (lighter grey shades in the floodplain represent lower ground, the estuary very dark).

Clearly, each flood outline can equate to a number of SLR and defence-raising scenario combinations. For example, the flood outline given by Figure 6 can represent a 1 m SLR with no defence raise, or a 2 m SLR and defences raised by 1 m or any other combination of defence-raising and SLR scenario that gives a net 1 m SLR. A total of 30 hydrodynamic simulations were performed for each scenario at net SLR increments of 0.2 m. By plotting contours of storm surge return period and net SLR, the appropriate flood outline can be selected; this is shown for Figures 6–8 in Figure 10. This figure also shows how even a 0.5 m SLR can significantly increase the probability of flooding. Thus, a rate of SLR of 1 m every 20 years will result in a significant increase in inundation probability for large areas of the Thames estuary after only 20–30 years, which is likely to pose significant challenges in terms of developing rapid adaptation strategies.

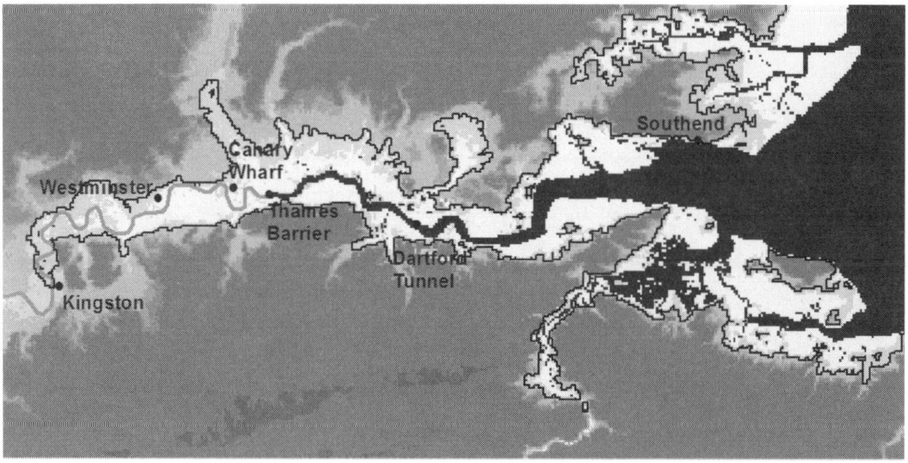

Figure 7. The 1:1000 year flood after 3 m net SLR.

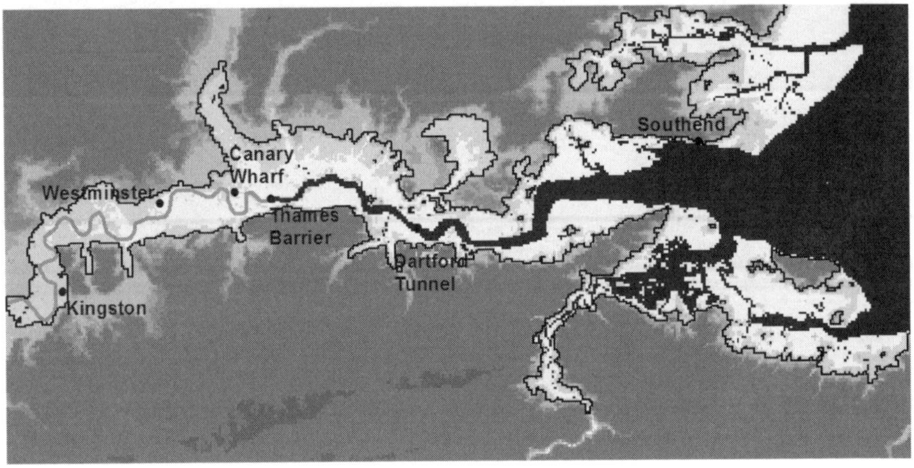

Figure 8. The 1:1000 year flood after 5 m net SLR, or 6 m SLR and 1 m defence upgrade (worst-case flood extent modelled).

Figures 4 and 5 show flood outlines for the Canvey Island and Sheppey barriers, respectively. The flood outlines are for a 1:1000 year flood event assuming 1 m SLR after the construction of barriers to a 1000 year standard of protection. The areas inundated for the outer barrier scenarios are 450 km^2 for a barrier located at Canvey Island and 350 km^2 for a barrier located at Sheppey Island, as compared with 650 km^2 for the existing Thames Barrier. Hence, construction of the outer barriers significantly reduces inundation probabilities in the floodplain. Their resilience to storm surges and high river flows is a result of the increased storage potential behind the barrier. This resilience, combined with their downstream position in the estuary, means that inundation of central London (i.e. Westminster) is much less likely, although East London (including Canary Wharf) would still be at risk (albeit considerably reduced).

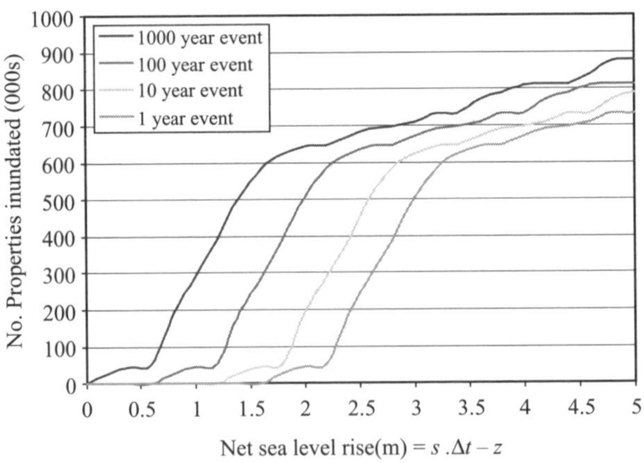

Figure 9. Number of properties inundated for different return periods and net SLR (= rate of SLR × $\Delta t - \Delta z$ where Δt equals SLR duration and Δz is the change in crest level height) based on current defence system and assuming no development in the floodplain after 2003.

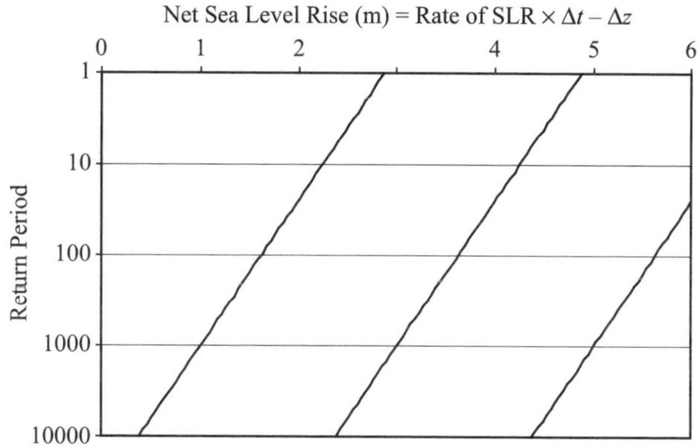

Figure 10. Contour lines showing how Figures 6–8 equate to a multitude of flood events and crest level-raising scenarios depending on the net SLR (= rate of SLR × Δt − Δz where Δt equals SLR duration and Δz is the change in crest level height).

Preliminary Consideration of Socio-economic Impacts

Grids providing spatial information of flood depth from the inundation modelling can be combined with flood depth–damage curves (e.g. Penning-Rowsell *et al.*, 2003) to estimate expected primary direct damages from flooding. For a given grid cell the expected annual damage, R, is given by:

$$R = \int_0^{y_{max}} p(y)D(y)\mathrm{d}y \tag{3}$$

where y_{max} is the flood depth that results in maximum possible damage to a property, $p(y)$ is the probability density function for flood depth and $D(y)$ is the damage at depth y. The total expected annual damage for the Thames estuary floodplain is obtained by summing the expected annual damage for each inundated raster cell. For the most extreme net SLR scenario of 5 m (Figure 8), during the 1000 year event, the economic damage is estimated (in terms of 2003 prices) at £39.6 billion for residential property and approximately £58.2 billion for non-residential property. These estimates do not include the economic damage to infrastructure such as the Underground rail system and, because they are based on national averages, may require weighting to account for higher property prices in the London region.

The high property density in the area will result in very high primary direct damages from such flood events. However, important commercial districts such as Canary Wharf are in the higher-risk zones—being placed under threat from only a small (~1 m net) increase in sea level. Whilst the direct damage to infrastructure would undoubtedly be large, the potential impact on the London and UK economy in the long term might be disastrous. The interruption to transport and communication links and the loss of working hours and trading, etc. during the inundation and subsequent clean-up could run into billions of pounds sterling. Given the dynamic nature of financial businesses located in London they might rapidly relocate to other financial districts in Europe, such as Frankfurt, that are perceived to be 'safer' in the long term. The international nature of business would mean the impact of a serious flood event is likely to have global economic repercussions (Munich Re, 2004).

Due to the extreme nature of the scenarios being considered it is likely that large areas of land are likely to be inundated rapidly. Water velocities cannot be reliably extracted from the inundation model, although rates of rise can be estimated and the model shows rates of rise in central London of up to 2 m/h for the 1:1000 year event after just 1 m net SLR. This rate of rise poses a serious threat to human life, particularly when considering that the high population density of the Thames estuary means that evacuation of large areas of floodplain poses a serious problem for the emergency services. The risk of inundation of a large number of London hospitals (and more elsewhere along the estuary) adds further to this strain. Supplies of drinking water, gas and electricity might be interrupted. Furthermore, inundation of heavy industry at Canvey Island and elsewhere could result in the spreading of contaminants and consequential environmental damage or conflagration as imagined by Doyle (2003).

These impacts are estimated based on the assumption that the socio-economic *status quo* is maintained and there is no further development in the floodplain. This scenario is unrealistic. First, significant development is already planned in the estuary, meaning impacts are likely to increase (ODPM, 2004). Secondly, the impact assessment is heavily simplified as it only considers direct damage resulting from independent flood events. Figure 9 demonstrates that even for small increases in net sea level, a large number of properties rapidly become susceptible to high probabilities of flooding. This increased flood frequency associated with the extreme SLR would result in more complex socio-economic feedbacks: society is likely to try and adapt to the SLR perhaps through increased provision of physical defences or even abandonment of frequently flooded areas. Consideration of these detailed socio-economic issues is outside the scope of this paper, but is reported by Lonsdale *et al.*, (forthcoming). However, this preliminary analysis is useful because it highlights the magnitude of the challenge that such an extreme SLR would pose to flood risk managers.

Limitations and Directions for Further Research

The current model has been implemented with relatively limited data. This is a major advantage of using LISFLOOD-FP and a key reason in its selection for this study. The model has demonstrated its power as an exploratory tool to assess flood impacts, including the benefits of possible interventions on flood risk. As well as defence-raising scenarios, two possible outer barrier sites have been tested—at Canvey Island and Sheppey—to provide an indication of how two alternative barrier positions, when compared to the present barrier, might reduce flood risk along the estuary. These demonstrate increased effectiveness at reducing flood risk. However, it should be noted that not all flood risk management adaptations have been tested and the authors do not wish the analysis of only 'hard' adaptation options in this paper to be misinterpreted as an expression of support for the construction of an outer barrier. Future developments of the analysis should consider managed realignment within the estuary, and other 'soft' flood management solutions such as those proposed by Hall *et al.*, (2003b), which include flood resilient development, public education and flood warning. Additional modelling would be required in order to simulate the effectiveness of an integrated portfolio of responses. Fluvial flood events are also likely to become more important given the possibility of a 30% increase in winter precipitation by 2080 (Hulme *et al.*, 2002), and the consequent increase in water run-off in the Thames (Reynard *et al.*, 2001). Further enhancement of the model would enable the flood risk associated with joint fluvial and tidal events to be considered.

Uncertainties in the inundation modelling can be quantified (Aronica *et al.*, 2002), although in this study, due to the extreme nature of the events being considered, uncertainties in the flood outline and depth will be dominated by the inaccuracies of the DEM. However, these are likely to be insignificant by comparison to uncertainties related to our poor understanding of the underlying climate processes. A number of possible mechanisms for abrupt climate change have been identified (e.g. collapse of the WAIS (Oppenheimer, 1998; Vaughan & Sponge, 2002), or collapse of the North Atlantic Thermohaline Circulation (Clark *et al.*, 2002)) but our poor understanding of the climate system and its sensitivity to anthropogenic forcing presently precludes the detailed estimation of scenario probability. Whilst the authors are not suggesting that the full range of modelled scenarios need to be explicitly considered in future design, the model results highlight the vulnerability of the Thames estuary floodplain to even 1 m net SLR. In order to respond to these poorly defined risks effectively, future flood management strategies need to be robust and consist of a portfolio of possible adaptation options so that they can respond to a range of events, including those induced by abrupt or large climate change. In the case of London, it can be argued that this range extends to a 2 m SLR during the 21st century (Arnell *et al.*, 2005).

This becomes increasingly important when the rate of change is considered. Should 1 m net SLR occur over a long duration, society is likely to be able to respond on an *ad hoc* basis. However, rapid change such as the partial or complete collapse of the WAIS would force a step change in the rate of SLR. Based on the experience of constructing the current Thames Barrier and the Eastern Schelde Barrier in the Netherlands, it is unlikely that a new outer barrier could be designed and built in less than 20 years. Indeed, even an upgrade of the current defence system is likely to require a similar time-scale. In an SLR scenario of 5 m over 100 years—even including the proposed upgrade of 1 m to the defence system—within 40 years central London will be threatened by the 1 in 10 year flood (Figure 9). Given that it may also take a number of years before data collection and analysis confirm that a step change in the rate of SLR has occurred, rather than it being identified as natural variability, it is possible that radical non-structural solutions, such as large scale abandonment, might be forced upon the inhabitants of the Thames estuary. This is echoed by Mastrandrea & Schneider (2001), who state that the advent of abrupt climate changes would reduce adaptability and thus increase climate damage. They are supported in arguing for stronger and earlier climate mitigation action by Perrings (2003), who believes that such mitigation can be justified economically by the risks of abrupt change. Given the great uncertainty in our understanding of human and climate related processes, future decisions on flood management need to be flexible in the long term in order to effectively manage this uncertainty (cf. Evans *et al.*, 2004).

Conclusions

The Environment Agency and other stakeholders are currently reviewing the long-term strategy for managing the Thames estuary with the aim of protecting London into the next century. This plan is based around preparing for the high-end UKCIP scenario for SLR of about 1 m over the 21st century. However, a greater rise in SLR is certainly possible: the collapse of the WAIS is one causal mechanism which might force a stepped change in the rate of SLR. While this paper has explored a range of extreme SLR scenarios, an important result is that a net SLR as low as 1 m would significantly increase flood risk in the Thames estuary. Given the likely 1 m upgrade of the defences, this translates into a 2 m rise in sea level. This could be produced by a variety of combinations of sea level sources during the 21st century, such as (1) the high-end IPCC (2001) scenarios (~0.9 m), (2) positive regional changes in the north-west Atlantic (~0.4 m) and (3) the balance from Antarctica (~0.7 m) (see Arnell *et al.*, 2005).

Modelling presented in this paper has for the first time linked a possible mechanism for abrupt climate change and SLR to a two-dimensional hydrodynamic model that can be used to support quantified impacts assessment. The model demonstrated that a stepped change in SLR could have disastrous consequences along the Thames estuary, including scenarios of partial collapse of the WAIS. A limited number of adaptations were tested and this demonstrated the effectiveness of an outer barrier as one approach to reduce flood risk under these conditions. A number of further improvements to the model have been identified, in particular the need to model non-structural solutions and consider fluvial–marine interactions. Most importantly, the results show the need for future flood management to be both robust and adaptive so that it can better respond to the range of possible changes, such as rapid SLR.

This paper has provided a quantified analysis of a small subset of possible responses under idealized conditions to extreme SLR scenarios, providing an important baseline for flood risk managers. The model results can be used to support detailed consideration of societal response. Although outside the scope of this paper, this is being investigated in a series of other papers from the Atlantis research project, including Lonsdale *et al.*, (forthcoming).

Acknowledgements

The research in this paper was funded through the European Commission FP5: EVK2-CT-2002-00138 as the Atlantis project (Atlantis: North Atlantic Sea-level Rise: Adaptation to Imaginable Worst Case Climate Change). The DEM was provided by the Environment Agency of England and Wales.

References

Arnell, N. W., Cannell, M. G. R., Hulme, M., Kovats, R. S., Mitchell, J. F. B., Nicholls, R. J., Parry, M. L., Livermore, M. T. J. & White, A. (2002) The consequences of CO_2 stabilisation for the impacts of climate change, *Climatic Change*, 53, pp. 413–446.

Arnell, N., Tompkins, E., Adger, N. & Delaney, K. (2005) Vulnerability to abrupt climate change in Europe. Available at www.tyndall.ac.uk, accessed October 2005.

Aronica, G., Bates, P. D. & Horritt, M. S. (2002) Assessing the uncertainty in distributed model predictions using observed binary pattern information within GLUE, *Hydrological Processes*, 16, pp. 2001–2016.

Bates, P. D. & De Roo, A. P. J. (2000) A simple raster-based model for flood inundation simulation, *Journal of Hydrology*, 236, pp. 54–77.

Bates, P. D., Dawson, R. J., Hall, J. W., Horritt, M., Nicholls, R. J., Wicks, J. & Hassan, M. A. A. M. (2005) Simplified two-dimensional numerical modelling of coastal flooding and example applications, *Coastal Engineering*, 52(9), pp. 793–810.

Chadwick, A. & Morfett, J. (1993) *Hydraulics in Civil and Environmental Engineering*, 2nd edn (London: E. & F. N. Spon).

Clark, P. U., Pisias, N. G., Stocker, T. F. & Weaver, A. J. (2002) The role of the thermohaline circulation in abrupt climate change, *Nature*, 415, pp. 863–869.

Colemand, M. D. & Mercer, J. B. (2002) NEXTMap Britain: completing phase 1 of Intermap's global mapping strategy, *GeoInformatics*, December, pp. 16–19.

Cugier, P. & Le Hir, P. (2002) Development of a 3D hydrodynamic model for coastal ecosystem modelling. Application to the plume of the Seine River (France), *Estuarine and Coastal Shelf Science*, 55, pp. 673–695.

Dawson, R. J. (2003) Performance-based management of flood defence systems. PhD Thesis, University of Bristol.

Dawson, R. J., Hall, J. W., Sayers, P. B. & Bates, P. D. (2003) Flood risk assessment for shoreline management planning, in: R. G. McInnes (Ed.) *Proceedings of the International Conference on Coastal Management*, pp. 93–97 (London: Thomas Telford).

Doyle, R. (2003) *Flood* (London: Arrow).

Dyer, K. R. (1973) *Estuaries: A Physical Introduction* (Chichester: Wiley).

Environment Agency (2003a) *Review and Update of the 'Tidal Thames Strategy Review'* (Bristol: Environment Agency).

Environment Agency (2003b) Thames *Gateway and Flood Risk Management—A Preliminary Assessment* (Bristol: Environment Agency).

Environment Agency (2003c) *Planning for Flood Risk Management in the Thames Estuary* (Bristol: Environment Agency).

Evans, E., Ashley, R., Hall, J., Penning-Rowsell, E., Saul, A., Sayers, P., Thorne, C. & Watkinson, A. (2004) *Foresight: Future Flooding. Scientific Summary: Volume 1—Future Risks and Their Drivers and Foresight: Future Flooding. Scientific Summary: Volume 2—Managing Future Risks* (London: Office of Science and Technology).

Gilbert, S. & Horner, R. (1984) *The Thames Barrier* (London: Thomas Telford).

Hall, J. W., Dawson, R. J., Sayers, P. B., Rosu, C., Chatterton, J. B. & Deakin, R. (2003a) A methodology for national-scale flood risk assessment, *Water and Maritime Engineering*, 156, pp. 235–247.

Hall, J. W., Meadowcroft, I. C., Sayers, P. B. & Bramley, M. E. (2003b) Integrated flood risk management in England and Wales, *ASCE Natural Hazards Review*, 4(3), pp. 126–135.

Horritt, M. S. & Bates, P. D. (2001) Predicting floodplain inundation: raster-based modelling versus the finite-element approach, *Hydrological Processes*, 15, pp. 825–842.

Hulme, M. (2003) Abrupt climate change: can society cope?, *Philosophical Transactions of The Royal Society A: Mathematical, Physical and Engineering Sciences*, A361(1810), pp. 2001–2021.

Hulme, M., Jenkins, G. J., Lu, X., Turnpenny, J. R., Mitchell, T. D., Jones, R. G., Lowe, J., Murphy, J. M., & Hassell, D., Boorman, P., McDonald, R. & Hill, S. (2002) *Climate Change Scenarios for the United Kingdom: The UKCIP02 Scientific Report.* (Norwich: Tyndall Centre for Climate Change Research, University of East Anglia).

IPCC (2001) *Third Assessment Report and Summary for Policy Makers* (Cambridge: Cambridge University Press).

Jones, D. A. (2001) *Results from the Joint Probability Analysis: Appendix B of Tidal Thames Probability Analysis.* (Wallingford: Centre for Ecology & Hydrology).

Kashefipour, S. M., Lin, B., Harris, E. & Falconer, R. A. (2002) Hydro-environmental modelling for bathing water compliance of an estuarine basin, *Water Research*, 36, pp. 1854–1868.

Keller, K., Tan, K., Morel, F. M. M. & Bradford, D. F. (2000) Preserving the ocean circulation: implications for climate policy, *Climatic Change*, 47, pp. 17–43.

London Assembly (2002) Flooding in London. Available at http://www.london.gov.uk/assembly/reports/, accessed October 2005.

Lonsdale, K. G., Downing, T. E., Nicholls, R. J., Parker, D., Vafeidis, A. T., Dawson, R., & Hall, J., (forthcoming) Plasuible responses to the threat of rapid sea-level rise for the Thames estuary, *Climatic Change, 5 (in press)*

Mastrandrea, M. D. & Schneider, S. H. (2001) Integrated assessment of abrupt climatic changes, *Climate Policy*, 1, pp. 433–449.

Mercer, J. H. (1978) West Antarctic Ice Sheet and CO2 greenhouse effect: a threat of disaster, *Nature*, 271, pp. 321–325.

Munich, Re (2004) Megacities—megarisks. Available at http://www.munichre.com/publications/302-04271_en.pdf, accessed October 2005.

National Research Council (2002) Abrupt climate change: inevitable surprises. Available at http://www.nap.edu/books/0309074347/html/, accessed October 2005.

Nicholls, R. J. (1995) Coastal megacities and climate change, *GeoJournal*, 37, pp. 369–379.

Nicholls, R. J. (2002) Rising sea levels: potential impacts and responses, in: R. Hester & R. M. Harrison (Eds) *Global Environmental Change: Issues in Environmental Science and Technology*, 17, pp. 83–107 (Cambridge: Royal Society of Chemistry).

Nicholls, R. J. & Lowe, J. A. (2004) Benefits of mitigation of climate change for coastal areas, *Global Environmental Change*, 14, pp. 229–244.

Nicholls, R. J., Hoozemans, F. M. J. & Marchand, M. (1999) Increasing flood risk and wetland losses due to global sea-level rise: regional and global analyses, *Global Environmental Change*, 9(S1), pp. S69–S87.

ODPM (2004) Greening the gateway. Available at http://www.odpm.gov.uk/, accessed October 2005.

Oppenheimer, M. (1998) Global warming and the stability of the West Antarctic Ice Sheet, *Nature*, 393, pp. 325–332.

Parry, M. L. (Ed.) (2004) An assessment of the global effects of climate change under SRES emissions and socio-economic scenarios, in: *Global Environmental Change*, 14, pp. 1–99.

Penning-Rowsell, E. C., Johnson, C., Tunstall, S. M., Tapsell, S. M., Morris, J., Chatterton, J. B., Coker, A. & Green, C. (2003) *The Benefits of Flood and Coastal Defence: Techniques and Data for 2003* (London: Flood Hazard Research Centre, Middlesex University).

Perrings, C. (2003) The economics of abrupt climate change, *Philosophical Transactions of The Royal Society A: Mathematical, Physical and Engineering Sciences*, A361(1810), pp. 2001–2021.

Reynard, N. S., Prudhomme, C. & Crooks, S. M. (2001) The flood characteristics of large UK rivers: potential effects of changing climate and land use, *Climatic Change*, 48, pp. 343–359.

Rossiter, J. R. (1954) The North Sea storm surge of 31st January and 1st February 1953, *Philosophical Transactions of The Royal Society A: Mathematical, Physical and Engineering Sciences*, A246(915), pp. 317–400.

Small, C. & Nicholls, R. J. (2003) A global analysis of human settlement in coastal zones, *Journal of Coastal Research*, 19, pp. 584–599.

Smith, K. & Ward, R. (1998) *Floods: Physical Processes and Human Impacts* (Chichester: Wiley).

Tirpak, D., Ashton, J., Dadi, Z., Filho, L. G. M., Metz, B., Parry, M., Schellnhuber, J., Yap, K. S., Watson, R. & Wigley, T. (2005) Report of the Steering Committee: International Symposium on the Stabilisation of Greenhouse Gases. Available at http://www.stabilisation2005.com/, accessed October 2005.

Tol, R. S. J., Bohn, M., Downing, T. E., Guillerminet, M.-L., Hizsnyik, E., Kasperson, R., Lonsdale, K., Mays, C., Nicholls, R. J., Olsthoorn, A. A., Preifle, G., Poumadereh, M., Toth, L., Vafeidisk, N., Werff, P. E. Vander & Yetkiner, I. H. (forthcoming) Adaptation to five metres of sea level rise, *Journal of Risk Analysis* (in press).

Turner, R. K., Subak, S. & Adger, W. N. (1996) Pressures, trends, and impacts in coastal zones: interactions between socioeconomic and natural systems, *Environmental Management*, 20, pp. 159–173.

Vaughan, D. G. & Sponge, J. (2002) Risk estimation of the collapse of the West Antarctic Ice Sheet, *Climatic Change*, 52, pp. 65–91.

Wigley, T. M. L. (1995) Global-mean temperature and sea level consequences of greenhouse gas stabilization, *Geophysical Research Letters*, 22, pp. 45–48.

Advances in Decision Support Systems for Flood Disaster Management: Challenges and Opportunities

JASON K. LEVY, CHENNAT GOPALAKRISHNAN & ZHAOHUI LIN

Introduction

Flood disasters have an enormous impact on human well-being, jeopardizing important social development goals such as addressing poverty, ensuring adequate food, water, and sanitation, and protecting the environment. In many regions of the world, flooding is a frequent, widespread, devastating, and increasing natural hazard. For example, in recent years, water levels in England (Ouse) and the European Continent (Elbe) rose to their highest levels in recorded history, breaking previous records set in 1625 and 1845, respectively. According to the latest statistics from the Office of Foreign Disaster Assistance and the World Health Organization's Collaborating Centre for Research on the Epidemiology of Disasters, during the last decade of the 20th century floods killed approximately 100 000 people, and affected over 1.4 billion (Jonkman, 2005). Direct losses from floods include drownings and severe injuries (particularly in flood-prone countries such as Bangladesh, where flood disasters in 2004 forced millions to go without adequate food or clean water) as well as damage to infrastructure and property, agricultural production, and sites of historical and cultural value (Appleton, 2002). Indirect health problems often arise, such as water-borne infections, exposure to chemical pollutants released into flood waters, vector-borne diseases, and a lack of clean water (Hajat *et al.*, 2003). With the exception of floods generated by dam failure or landslides, floods are climatological phenomena and hence climate variability and global climate change play a significant role.

Since 1850, average global temperatures have risen about 0.6°C, most likely due to anthropogenic activities such as the burning of fossil fuels that elevate the ambient concentration of greenhouse gases. Global climate change implies that heat over the Earth's surface will be redistributed (some areas will warm, others will cool). The United Nations' (UN's) third assessment report of the Intergovernmental Panel on Climate Change (IPCC) (2001) projects that global temperatures will rise between 1.4°C (2.5°F) and 5.8°C (10.4°F) by 2100. A warmer climate, and the accompanying shifts in rainfall patterns, are expected to lead to more rainfall (and higher flood risk) in many parts of the world, particularly in maritime areas such as north-west Europe. Some changes are already evident: throughout the 20th century, precipitation in the USA increased nationally by about 5–10%, mainly due to an increase in intense rainstorms (National Assessment Synthesis Team, 2000). In England, rainfall in the autumn of 2000 exceeded all previous records since precipitation records generally began in 1776 (National Audit Office, 2001).

In addition, IPCC (2001) climate scenarios project an increase in tropical cyclone activity. Tropical cyclone induced storm surges constitute one of the most significant natural disaster risks in terms of property devastation and lives lost (Zerger *et al.*, 2002). As eloquently summarized by UN Secretary General Kofi Annan in the wake of the December 2004 Indian Ocean tsunami disaster, "it is no longer so hard to imagine what might happen from the rising sea levels that the world's top scientists are telling us will accompany global warming" (Annan, 2005). By 2100 global sea levels are expected to rise another 4 inches (1 inch = 2.54 cm) due to thermal expansion (physical swelling) of the oceans as they warm (Meehl *et al.*, 2005). When including fresh water that is expected to melt from ice sheets and glaciers, sea levels are expected to rise by as much as 10 inches per century until the year 2400 (Wigley, 2005). Even if greenhouse gas concentrations are stabilized, a significant ongoing large sea level rise is expected to continue over the next century (and possibly far beyond) because oceans have an enormous capacity to absorb heat and react relatively slowly to changes in atmospheric and surface temperature conditions.

In addition to hydro-meteorological climate variability and change, a number of factors are increasing the vulnerability of society to coastal and freshwater flooding events, including land-use changes (such as the alteration of vegetation along rivers and coastlines) and urbanization, particularly the development of infrastructure in floodplains. For example, the 1998 Yangtze floods in China caused more than 3600 deaths and US$30 billion in economic losses due to the heavy concentration of infrastructure in the Yangtze River valley (Wei *et al.*, 2004). Floodplains throughout the UK are densely developed, containing approximately 1 million residential, commercial, and industrial properties worth over US$150 billion (Hall *et al.*, 2003). In Japan, half of the population and three-quarters of its property are densely concentrated in flood-prone alluvial plains. In the Netherlands, 7.5 million people and 50% of the country is at risk of flooding (National Audit Office, 2001). Last year in Bangladesh, floods submerged over two-thirds of the country.

Accordingly, flood disaster prevention and vulnerability reduction are topics of major concern for local, national, and international organizations. The UN declared the last decade of the 20th century to be the International Decade for Natural Disaster Reduction. In the following years, the UN world disaster reduction campaigns have attempted to heighten awareness of socio-economic vulnerability to floods. Despite these efforts, many nations remain vulnerable to floods. For example, in the USA, flood management practices have evolved significantly over the past half-century, and now include detailed flood hazard maps, advanced hydrologic models, early warning systems, and evacuation procedures. However, over this period, flood-related deaths have remained relatively constant, averaging 70 or more in each of the past five decades (Mileti, 1999). On the other hand, economic damage from US floods has increased significantly, climbing to approximately US$6 billion per year in 2005 (American Water Resources Association, 2005).

To cope with these challenges, a comprehensive flood decision support system (DSS) architecture is proposed. Flood DSSs constitute a customized, interactive computing environment (with interlinked models/analytical tools, databases, graphical user interfaces, and other systems) designed to help hydrologists, stakeholders, government officials, and other decision makers use data and models to evaluate flood risk management problems. Several international initiatives have recently emphasized the importance of DSS development for flood planning and management, including the European Union's River Basin Modelling Management and Flood Mitigation project, Agenda 21 (United Nations Conference on Environment and Development), and the European Union's Sixth Environment Action Programme. Flood DSSs have the potential to mitigate flood disasters though improved data collection and the rapid dissemination of information to affected areas, such as maps of potential flood water distribution. Improved flood DSSs are expected to increase the capacity of society to anticipate, resist, cope with, and recover from the impact of a flooding disaster.

Flood Risk Management under a Changing Climate: Hard vs. Soft Approaches

Climate change is one of the most complex and important issues of modernity, both in terms of the decisions needed to mitigate it and the actions that will be necessary to adapt to a changing climate. Since the Industrial Revolution, humans have been responsible for releasing ever-increasing amounts of carbon dioxide and other greenhouses gases into the atmosphere. Over the past quarter-century, average surface temperatures in the Arctic Circle have risen by more than 0.5°C (0.9°F) per decade. Rising global temperatures are lengthening melting seasons, thawing frozen ground, and thinning ice caps and glaciers. These changes are raising sea level faster than expected, dramatically increasing future flood risk. Specifically, the Arctic melt season has lengthened by approximately 2 weeks and the extent of Arctic sea ice cover has decreased by nearly 10% a decade (there have been record small extents of summer Arctic ice in recent years). The melting of Greenland's ice sheet (the largest land ice mass in the Northern Hemisphere) is another serious concern. The melting of this ice sheet would release sufficient fresh water to raise the Earth's sea level by 7.2 m (24 feet). Increasing concentrations of greenhouse gases and rising regional temperatures have caused Greenland's ice to melt at increasingly higher elevations: annual ice loss from Greenland is estimated at 50 km^3 (12 cubic miles) per year, enough water to raise the global sea level by 0.13 mm/year (Murray, 2005). As ice shelves in Antarctica, Greenland, and other regions collapse, the barrier to the sea for interior glaciers is removed, leading to a more rapid loss of fresh water to the ocean. As noted by Murray (2005), "glaciers in West Antarctica are discharging about 250 cubic kilometers of ice and water into the ocean per year, 60% more than is accumulated in their catchment areas—a net change sufficient to raise global sea levels by more than 0.2 millimeters per year". It is important to note that many of the effects of global warming are self-reinforcing: positive feedback could lead to undesirable climate change outcomes and significant increases in flood risk. For example, thawing tundra releases trapped carbon dioxide and methane from the newly formed wetlands. This further adds to increasing global temperatures. Finally, the ocean's capacity to act as a carbon sink may be compromised as waters warm.

Climate change and rising sea levels are expected to increase the intensity and frequency of flood events around the world. For example, flooding on the Mississippi and Missouri rivers has substantially increased over the past century: Pinter (2005) and others have noted that dramatic increases in flood levels have been observed over the past 100 years along the lower Missouri River and the lower Mississippi River near St Louis (increases of up to 3–4 m). Economic losses from flood disasters are also on the rise. The great US Mid-western flood of 1993 affected over 1600 km of the Mississippi and Missouri rivers and caused up to US$16 billion in damages (Myers & White, 1993). Both before and after this flood, a number of formal flood control recommendations emerged in the USA to reduce flood losses, such as constraining (or even reducing) floodplain infrastructure. To this end, the Federal Emergency Management Agency (FEMA) has bought out over 7700 properties in Illinois and Missouri (two states heavily affected by floods) at a cost of US$56.3 million, including the relocation of an entire Illinois town. Unfortunately, these valuable management decisions are compromised by the construction of new properties and infrastructure in floodplains (Pinter, 2005).

The St Louis area provides a poignant example of how local economic interests and political pressure can increase flood risks by increasing floodplain development. Since the 1993 great Mid-western flood, over 28 000 new homes and 26.8 km^2 (6630 acres) of commercial and industrial developments have been constructed on land that was under water during the 1993 flood event (Pinter, 2005). Shockingly, in the St Louis area alone, this constitutes over US$2 billion in new construction in formerly flooded regions. To protect this new infrastructure from floods, since 1993, over 72.8 km^2 (18 000 acres) of the Mississippi and Missouri floodplains near St Louis will soon be behind levees. Many levee construction projects receive support from local governments and the US Army Corps of Engineers, which has spent nearly US$200 million on levees in the St Louis district since 1993. While FEMA guidelines under the National Flood Insurance Program limit development in the floodway (central portion of the floodplain), construction in remaining floodplain areas is virtually unconstrained as long as the development occurs in areas that are either higher than the 100-year flood level (the event with a 1% chance of occurring in any year) or protected by levees with at least 100-year protection (Pinter, 2005).

The hard (structural) approach not only requires significant government expenditures (for new and enlarged structures to protect subdivisions and infrastructure build in floodplains) but also leads to adverse social and ecological effects. For example, humans continue to extensively alter river systems for flood control, water, energy, and transportation needs. As discussed by Nilsson *et al.* (2005), there are now over 45 000 dams above 15 m high and more than 300 giant dams which meet one of three criteria on height (over 150 m), dam volume (over 15 million m^3), or reservoir storage (more than 25 km^3). The catchment scale impacts of dams on ecosystems and humans are well known: fragmentation and flow regulation of the world's large river systems have led to significant upstream and downstream effects (obstructing the dispersal and migration of organisms, hindering channel development, modifying aquatic communities, etc.).

Moreover, levee enclosed areas are not immune from flooding events. As clearly articulated by a panel of the National Academy of Science, "it is short-sighted and foolish to regard even the most reliable levee system as fail-safe" (National Research Council, 1982). This panel also noted that one-third of all flood disasters in the USA were caused by levee failures. In addition, there are a host of legal issues associated with flood and levee hazards. Specifically, governments which encourage a programme of flood control levees may be held legally responsible for "any flood damages that result from improper design or maintenance of such systems" (National Research Council, 1982). In recent years, municipal, county, and state governments have been held increasingly liable for flood damages in instances where floodplain encroachment was encouraged and natural flooding patterns were modified with flood control systems.

The overall effectiveness of the hard approach ('keep floods away from people') has been called into question in light of the aforementioned social, legal, and economic concerns. Overall, flood control structures may actually increase damage potential by encouraging development along flood-prone waterways. While the USA continues to rely heavily on structural flood control measures, other countries have taken a more proactive, 'non-structural' (soft) approach to reduce flood damage ('keeping people away from floods'). The soft approach uses a variety of measures to reduce the vulnerability of existing population centres, infrastructure, and property. These include altering upstream land management practices, floodplain mapping and zoning (to ensure that new development in the floodplain does not occur), proactive disaster planning and management (evacuation systems, emergency response, etc.) and flood-proofing (e.g. elevating structures, sealing openings). This paper discusses how advances in DSSs can supplement the soft approach for flood risk management.

In recent years, European governments have shifted towards a strategy of soft flood risk management, recognizing that any flood containment structure will eventually be overcome. In particular, after the severe Rhine river flooding of 1993 and 1995, the Dutch government adopted a flood control policy of 'more room for rivers' with an emphasis on establishing new storage and conveyance space (the 'Room for the Rhine' doctrine was adopted in 1997). In this spirit, six European governments created the Meuse River High Water Action Plan, which provides "space for the river" and "longer storage and more release". Today, many European countries are providing significant financial resources for soft flood management approaches: the Netherlands alone has allocated US$3 billion for a broad array of levee alternatives. The soft approach has delivered impressive results. For example, in the state of Baden-Wurttemberg, Germany, the addition of over 200 m^3 of floodplain storage has reduced flood stages to 1950 levels. Some floodplain organizations in the USA, such as the Association of State Floodplain Managers, have also adopted the soft approach to minimize flood risks for humans and their property.

A DSS Architecture for Flood Disaster Management

Advances in flood DSSs can improve flood risk assessment, thereby providing a rational basis for the development of flood management policy at local, regional, and national scales (Hall *et al.*, 2003). Flood DSSs improve the decision-making process not by prescribing a particular course of action, but rather by providing data displays, analytical results, and model output to summarize critical flood information. Miller *et al.* (2004) outline four important challenges for developing effective DSSs related to information technology: ensuring the interoperability of technologies (including computer applications, platforms, and operating systems); providing the widest accessibility possible (with particular consideration to geographic information systems (GIS)–internet architecture, required plug-ins, and bandwidth); ensuring internet security; and addressing data ownership issues.

An overall architecture for flood DSSs is shown in Figure 1, with particular reference to flood reservoir management. It consists of three main components: a flood database, flood modelling functions; and a graphical user interface. The first DSS component is a flood database that typically includes meteorological, hydro-geologic, administrative, and population data to form a comprehensive depiction of a hydrologic region. Remote sensing from satellites provides another important source of data, often used in GIS. For example, the Hydrologic Information System's relational hydrologic observation database (Consortium of Universities for the Advancement of Hydrologic Science, 2004) contains observational data on stream flow, climate, water quality, and groundwater levels, together with remote sensing/GIS data and weather/climate grids. Advances in information technology such as fibre optic technology for river monitoring provide additional sources of data. In general, digital flood data are easy to process, highly accessible, and less expensive than traditional approaches to the delineation of floodplain boundaries.

The second component of a flood DSS is a suite of flood modelling functions, tailored to the client's needs. Flood models typically involve rainfall–run-off simulations, discharge levels, data exploration, and assessment. In addition, information on flood magnitudes and their frequencies is needed for the design of hydrologic structures, floodplain zoning, and the economic evaluation of flood management systems. Advances in soft computing technologies (fuzzy logic, artificial neural networks, and evolutionary computation), probabilistic predictions of inundation recurrence intervals (and magnitudes), and hydrodynamic flood models provide flood managers with a variety of powerful techniques for flood simulation, modelling, and management. Because flood processes are complex multivariate events characterized by correlated random variables (flood peak, volume, duration, etc.), multivariate flood modelling offers improved understanding and application over univariate hydrologic techniques (Kite, 1988; Stedinger *et al.*, 1993), but require considerably more data and sophisticated mathematical analysis.

Flood multiple-criteria decision-making (MCDM) models constitute a collection of methodologies to compare, select, or rank flood alternatives that involve incommensurate attributes. MCDM models are well suited to addressing flood management problems because decision makers must typically deal with large-scale (involving thousands of variables and competing alternatives) and complex (involving the simulation of run-off and the determination of water surface profiles) challenges. Many MCDM flood decisions are formulated as mathematical programming problems, where a function of a vector of decision variables is to be optimized while constrained to remain within a feasible region. The focus is on alternatives on the boundaries of this region, termed the efficient frontier (for example, alternatives Q, R, S, T, and W in Figure 1). Specifically, MCDM models can identify alternatives that are dominated by at least one other alternative (i.e. have poorer values on some criteria, and no better values on another criterion). In general, such dominated (inefficient) alternatives should not be considered further as they fall below the efficient frontier. For instance, in Figure 1, alternative P is dominated by alternatives Q, R, and W.

Finally, MCDM models are helpful for exploring challenging and value-laden flood management trade-offs. For example, in reservoir operation and management problems, flood protection (minimizing reservoir discharge during peak flood periods) and energy

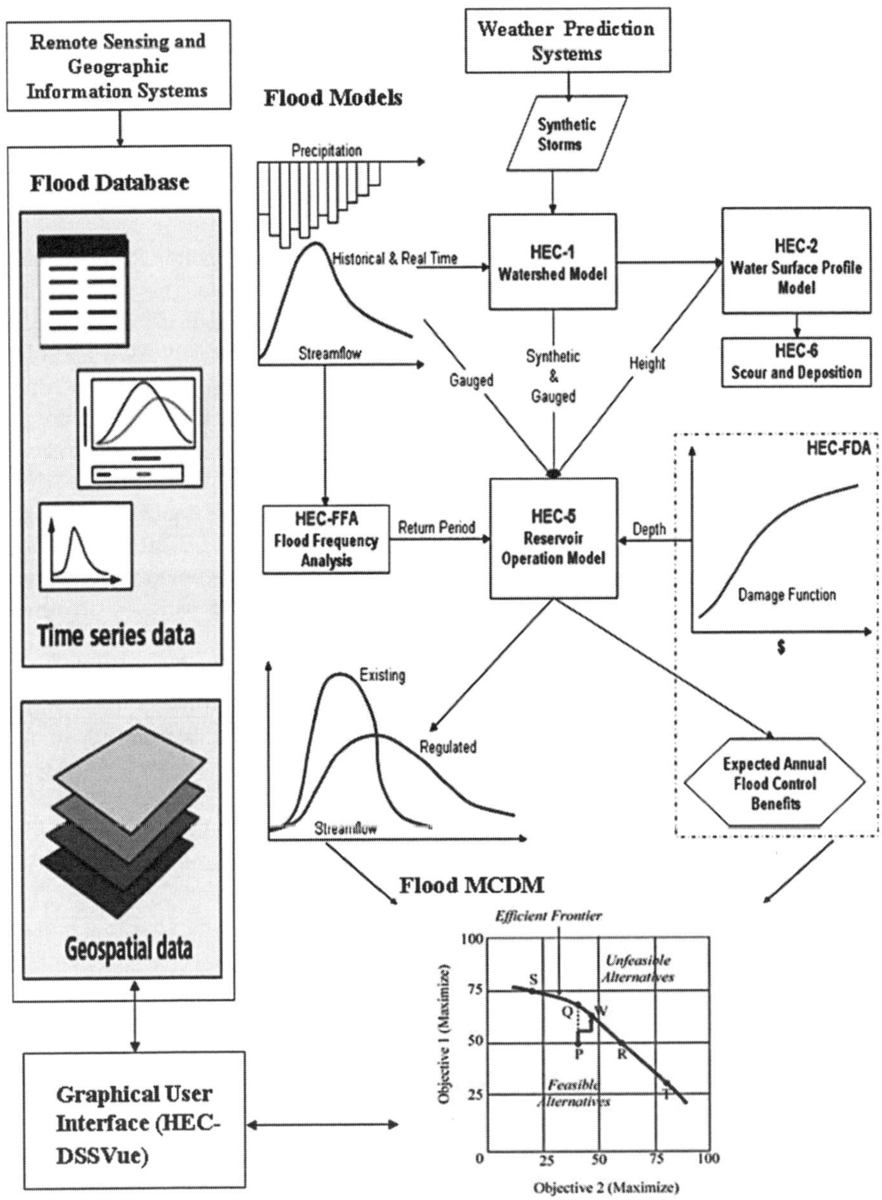

Figure 1. DSS architecture for flood reservoir operation: database, models, and a graphical user interface.

production (meeting a pre-defined energy production target) are important priorities. Flood protection implies that the reservoir should be kept at as low a level as possible, so that the reservoir can absorb the inflows during flood periods. On the other hand, the energy production objective requires as much water in the reservoir as possible.

The analytic network process (ANP) constitutes a popular MCDM approach (Saaty, 2004) With the ANP, one constructs hierarchies or feedback networks, and then makes judgements or performs measurements on pairs of elements with respect to a controlling element to derive relative absolute scales that are then synthesized throughout the structure to select the best alternative. The criteria are pairwise compared with respect to the goal, the subcriteria with respect to their parent criterion, and the alternatives of choice with respect to the last level of subcriteria above them (the covering criteria). Each such set of comparisons yields a relative absolute scale of priorities. The priorities throughout the network structure are then synthesized to give the overall priorities for the flood management alternatives under consideration.

Customized flood DSS modules are capable of solving spatially explicit flood modelling equations and addressing complex flood risk management problems such as flood evacuation emergency planning (Simonovic & Ahmad, 2005) and flood risk mitigation and control (Mysiak *et al.*, 2005). One of the best known and widely used hydrologic models is HEC-1 (flood hydrograph package) developed by the US Army Corps of Engineers' Hydrologic Engineering Center (HEC). First released in 1968, HEC-1 provides a variety of options for simulating rainfall–run-off processes. Figure 1 illustrates the interrelationships among a number of HEC models. For example, HEC-1 results can be used in HEC-2 (water surface profile model), the standard for FEMA's floodplain evaluations and river channel design. Next, output from HEC-1, HEC-2, HEC-FFA (flood-flow frequency analysis), and HEC-FDA (flood damage assessment) models can serve as input for HEC-5 (reservoir operation model). The output of HEC-5 includes stream flow and expected annual benefits accrued from flood control.

The third component of a flood DSS is a human–computer interface for interactive flood queries, reporting, and display. Frank (1993) notes that "the user interface is the system". DSS display capabilities include spatial data handling, the editing of flood inundation maps, sensitivity analyses of flood model parameters at a variety of spatio-temporal scales, analysis of complicated levee systems, and the animation of hydrologic and hydraulic phenomena. Interfaces determine the scope of the interaction between the computer system and the user (Al-Sabhan *et al.*, 2003). For example, HEC-DSSVue (Figure 1) is a Java-based graphical user interface program for viewing, plotting, editing, and manipulating data in HEC-DSS database files. Trade-offs among DSS functionality, efficiency, and required knowledge must be considered: interfaces should be customizable (adaptive) so that displays match the user's skill level (Kingston *et al.*, 2000).

Remote Sensing/GIS Methods

Over the past two decades, advances in remote sensing and GIS techniques have provided valuable new data for flood monitoring, flow forecasting, dam control, and river system management. Current remote sensing tools include weather radar, aircraft measurement, the detection of atmospheric electrical disturbances, and rain gauges. Remote sensing methods provide reliable and cost-effective field data collection, allowing for continuous, large-area, synoptic coverage of many variables (including snow melt, water level, ocean waves, surface winds, and sea surface height), although calibration with ground-measured data is required. For example, satellite monitoring provides relatively inexpensive, real-time estimates of rainfall (once a satellite is in orbit). There are a number of advantages of remote sensing techniques over ground-based methods. First, in many developing countries affected by floods, from Bangladesh to Mozambique, the density of gauging stations is quite low, leading to a shortage of ground data for flood risk assessment models (Sanyal & Lu, 2004). Second, ground stations may fail to accurately collect data in an extreme flood event. Third, multi-date satellite imageries can be used to monitor how human activities and natural phenomena affect land cover over time and to reconstruct the history of previous extreme hydrologic events. Recently, the TOPEX/Poseidon and Jason environmental satellites (joint USA–France altimetry missions) have been used to investigate the effects of ocean changes on climate events (such as El Niño and La Niña) and to analyse the relationships between anthropogenic activities and global change. On 20 May 2005, the NOAA-N weather satellite was launched from Vandenberg Air Force Base, California, in order to improve the monitoring of global climate changes. The fourth in a series of five polar-orbiting weather satellites, the 3100 lb. NOAA-N (1 lb. = 0.45 kg) will collect and transmit meteorological data in order to enhance databases on climate change and weather modelling. NOAA-N will replace the ageing NOAA-16, which has been operating since 2000. The last satellite in this series, NOAA-N Prime, is scheduled for a 2007 launch.

There are two major uses of remote sensing for flood research: as a tool for the delineation of flooded areas (detecting flood boundary river inundation, stage, and discharge); and as a technique for flood hazard and risk mapping. The pioneering uses of satellite remote sensing for monitoring floods and delineating the boundary of inundation used optical remote sensing, specifically data from the Landsat Multi Spectral Scanner (MSS) with 80 m resolution. The earliest applications involved flood-prone US regions, including the Mississippi River basin (Deutsch *et al.*, 1973; Rango & Anderson, 1974). By the early 1980s, the main source of data for flood monitoring and delineation was the Landsat Thematic Mapper (TM) imageries with 30 m resolution. MSS band 7 (0.8–1.1 μm) and Landsat TM NIR (near infra-red) band 4 proved valuable for distinguishing water (or moist soil) from dry land surface due to the strong absorption of water in the near infra-red range of the spectrum (Smith, 1997). For use in urban areas (industrial and commercial zones with asphalt), a combination of Landsat TM band 7 (2.08–2.35 μm) and band 4 was shown to successfully delineate inundated regions (Wang *et al.*, 2002).

At a larger scale, these medium-resolution imageries are often supplemented by more coarse resolution imageries such as remotely sensed Normalized Difference Vegetation Index (NDVI) data from the National Oceanographic and Atmospheric Administration's Advanced Very High Resolution Radiometer (AVHRR). AVHRR imageries enable the monitoring of flood progress in near real time (high temporal resolution). Limitations of AVHRR imageries include a coarse resolution (precluding local flood analysis), difficulties in selecting an appropriate NDVI threshold in order to distinguish inundated from dry surfaces, and sensitivity to satellite viewing angle and weather conditions (clouds are common during the monsoon season) (Barton & Bathols, 1989; Wang *et al.*, 2002). The development of microwave remote sensing, particularly radar images, permits the monitoring of flood progress during bad weather conditions because the radar pulse can penetrate cloud cover. An additional advantage of synthetic aperture radar (SAR) imagery is the ability to clearly distinguish between inundated and dry surfaces.

A combination of optical (Landsat TM imageries) and microwave remote sensing (SAR) has yielded superior results for dealing with challenging flood management problems such as flood mapping in mountainous areas: in mountains, most areas appear dark or shaded, which makes it difficult to identify flooded areas (only slopes perpendicular to the radar beam appear bright). However, using data obtained from different sensors remains challenging, often requiring iterative experiments and land cover maps to obtain better results. For example, land cover maps can minimize the risk of including permanent water bodies as flooded regions (Tholey *et al.*, 1997). In the context of a monsoon flood, wet season imagery is necessary: using imagery captured in the dry season would underestimate natural drainage, thereby overestimating flooded areas (Islam & Sadu, 2002).

Remotely sensed and hydrologic data are often used together with GIS to determine flood hazard potential and flood risk maps. For example, AVHRR data superimposed over a digital elevation map can facilitate the assessment of flood depth by using the tonal difference of the flood water and supervised classification (to subdivide the flood affected area into different flood depth zones). Land development priority maps (a function of flood hazard and population) are often created with remote sensing data and GIS for flood counter-measures (Islam & Sadu, 2002). The increasing availability of appropriate spatial data and the improved performance of computer technology have led to a growth in the use and application of GIS technology for analysing and visualizing spatial flood information, for empowering stakeholders, and for facilitating flood simulation, evacuation, and damage estimation. A temporal (time-integrative) GIS system (Ott & Swiaczny, 2001) captures spatial, temporal, and attribute data (such as flood height) and hence allows for the modelling of time-dependent flood processes and human activities, as well as the consideration of cross-correlations and dependencies in the composite spatio-temporal domain. Information about the timing and height of a flood peak, for instance, can greatly facilitate flood management decisions.

There are three broad approaches for integrating GIS and flood modelling (Clark, 1998): pre-processing data into a suitable format (such as spatial databases); direct GIS support for hydrologic models; and post-processing data. Examples of data processing in the pre-modelling phase include automated floodplain delineation from digital terrain models, raster-based models for flood inundation simulation, and the parameterization of flood models (Zerger & Wealands, 2004). The second approach involves direct GIS support (analysis, calibration, and other tasks) for flood modelling. Correia *et al.* (1999) argue that direct GIS support is often restricted to relatively simple hydrologic calculations due to proprietary data formats and algorithms. On the other hand, open architectures, such as those supported by the Open Geospatial Consortium, are beginning to address these challenges. The third approach for linking GIS and flood modelling involves post-processing data. These activities are becoming increasingly ubiquitous and typically include creating flood risk maps, performing cost–benefit analyses, and creating a flood decision support framework (Vermieran & Watson, 2001).

Seasonal Climate Prediction for Flood DSSs: Physical Basis and Modelling Approaches

Despite the chaotic nature of the atmosphere, there are a number of atmospheric components which exhibit variations on long time-scales (seasons and years) and which can, to some extent, be predicted: the ENSO (El Ni) cycle is the most important known source of year-to-year climate variability. Seasonal climate prediction attempts to predict the specific climate for forthcoming seasons and to distinguish how a predicted climate varies from the long-term climate (Stockdale, 2000). Seasonal rainfall prediction can improve flood planning and management activities such as flood evacuation, land-use planning, and reservoir operation. For example, the early warning of rainfall anomalies (with a lead time of one to several seasons) can advise managers to lower reservoir water levels before the onset of the rainy season, in addition to improving drought and crop management (e.g. Hammer *et al.*, 1996). In short, dynamic seasonal forecasting can reduce flood risk by providing valuable information about the climate that can be expected in coming months (or seasons), thereby helping to minimize human casualties and property damage.

The physical basis of seasonal predictability involves factors with relatively long time-scales of variability (relative to internal atmospheric variability) in the surface boundary conditions. This allows for the predictability of climate anomalies from one to several seasons in advance (far beyond the predictability limit of the atmosphere). Of particular interest are boundary conditions that vary slowly and therefore impose a slow variation on atmospheric states, such as sea surface temperature (SST), soil moisture content, and the extent of snow cover and sea ice: the large thermal inertia of the upper ocean and the persistence of land surface conditions can extend the predictability limit (of the atmosphere) to a few months and even longer (e.g. Charney & Shukla, 1981; Rosati *et al.*, 1997).

There are two general approaches for seasonal climate prediction: statistical/empirical models (hereafter referred to as statistical models) and dynamic/numerical models (hereafter referred to as dynamic models). Statistical seasonal prediction methods are based on historical data and employ a mathematical relationship between predicted and predictor variables. On the other hand, dynamic models derive seasonal predictions by solving the complex set of hydrodynamic equations that describe the evolution of the Earth's climate system. This involves using either stand-alone atmospheric general circulation models (AGCMs) or coupled global ocean–atmosphere circulation models (GCMs).

Statistical seasonal prediction methods have been in use since the beginning of the 20th century (e.g. Quayle, 1929). These techniques base seasonal predictions on the statistical analysis of past weather and climate. Statistical predictions have been highly successful in many regions of the world. For example, statistical techniques are able to capture the connection between rainfall in the Nordeste region of Brazil and SST in the tropical Atlantic Ocean. In many cases, SST anomalies (especially over the tropical Pacific Ocean) are the sole predictor for the statistical forecasts of seasonal climate anomalies (e.g. Folland *et al.*, 1991; Ward & Folland, 1991; Barnston, 1994). For example, the Southern Oscillation Index, based on SST, has been used for forecasting the seasonal climate of rainfall anomalies over Australia (e.g. Zhang & Casey, 1992; Stone *et al.*, 1996). In addition, atmospheric information can be an important factor for the analogue seasonal prediction technique (e.g. Livezey & Barnston, 1988; Lin *et al.*, 2003).

Seasonal climate prediction commonly uses both linear and non-linear statistical methods. Widely used linear statistical methods for climate prediction include regression approaches (e.g. Knaff & Landsea, 1997) and linear multivariate methods, such as canonical correlation analysis (CCA). For example, Barnston & Ropelewski (1992) apply CCA to predict El Niño variability, where seasonal-mean values of SST are used as the predicted variables (additional inputs include tropical Pacific SST, depth of the 20° isotherm, and sea level height over previous seasons). There are a number of limitations to statistical seasonal climate prediction. For example, statistical approaches require long and accurate data of the Earth's climate, although long-term observation data are often unavailable. Statistical seasonal forecasting techniques also require an understanding of the physically based relationships between predicted and predictor variables. In addition, all linear statistical models for seasonal prediction may suffer from instabilities in predictor–predictand relationships (e.g. Krishna *et al.*, 1999; Wu & Wang, 2002), leading to a decline in the skill of statistical seasonal predictions. Accordingly, many attempts have been made recently to apply non-linear statistical methods to seasonal climate prediction, such as the use of neural network models to forecast equatorial Pacific SST and to predict El Niño development (e.g. Tangang *et al.*, 1998; Sahai *et al.*, 2000).

Dynamic climate prediction methods constitute a valuable alternative to statistical approaches. Dynamic methods are based on a mathematical representation of physical laws governing the behaviour of the atmosphere or the coupled atmosphere–ocean system. There is a large body of research dealing with seasonal to inter-annual dynamic climate prediction (e.g. Bengtsson *et al.*, 1993; Ji *et al.*, 1994; Palmer & Anderson, 1994; Molteni *et al.*, 1996; Mason *et al.*, 1999). In addition, many international projects and collaborations have been undertaken to improve the skill of dynamic seasonal predictions (e.g. Palmer *et al.*, 2000). Dynamic seasonal climate prediction makes use of either a GCM or a stand-alone AGCM. A wide variety of atmospheric elements can be predicted using dynamic seasonal prediction because comprehensive physical processes have been parameterized within both GCMs and AGCMs.

The Earth's climate is a prototypical chaotic system. Accordingly, dynamic approaches for seasonal forecasting require a large number of separate realizations. If very large ensembles of climate predictions are conducted, one can make valuable inferences about the range of climate events which are most likely to occur during the next season (i.e. the probability distribution of seasonal climate anomalies). Specifically, ensemble prediction techniques have been proposed to generate probability forecasts by using a number of runs of a climate system model (Molteni *et al.*, 1996). Each run of the model typically differs by perturbations to initial conditions (and perhaps the model physics). However, there are a number of constraints that limit the dynamic seasonal forecast skill. For example, uncertainties related to the representation of physical processes within the model may adversely affect the performance of numerical atmospheric and ocean models. In addition, seasonal forecast skill may be significantly compromised if ocean observations are sparse.

The strengths and weaknesses of statistical and dynamic seasonal climate prediction methods are well documented. Currently, in seasonal forecasting, as in other areas of science, a combination of statistical and dynamic approaches often provides the most valuable seasonal prediction results. For example, in the context of monsoon prediction in India, Hastenrath (1995) found that the statistical method outperformed the dynamic approach. On the other hand, over southern Africa, Landman *et al.* (2001) noted that the seasonal forecast skill of dynamic models appeared to slightly outperform purely statistical approaches. However, as global climate models become more sophisticated (and their performance continues to improve) dynamic seasonal prediction has become more widely used. In particular, ensemble prediction techniques are capable of capturing the uncertainties (in initial conditions, parameterization schemes, etc.) related to seasonal climate prediction.

There have been significant improvements in real-time flood forecasting and warning systems in the past 10 years. The recent scientific and technological advances have been harnessed to increase the spatial and temporal resolution, accuracy, and forecast lead time for precipitation, run-off, and other hydrologic variables. In general, a 3-day weather forecast for precipitation is about as accurate today as a 1 day forecast was a decade ago. Werstedt & Hersh (2002) demonstrate the potential utility of climate forecasting for flood planning and management in the context of technical and institutional challenges. Numerical weather prediction (NWP) models describe atmospheric evolution through three-dimensional partial differential equations, based on physical laws such as the conservation of mass, momentum, and energy. The ultimate limit to the predictability of NWP models is determined by the projection of uncertainties in both initial atmospheric conditions and model formulation onto flow-dependent instabilities of the chaotic climate attractor (Lorenz, 1969). Ensemble prediction systems (EPSs), developed in the early 1990s, constitute an effective technique for improving weather and climate forecast skill: probability forecasts are generated by using a number of runs of an NWP model (differing by small perturbations to the initial conditions and model physics).

EPSs use dynamically conditioned perturbations in order to quantify the uncertainty in medium-range forecasts arising from synoptic-scale baroclinic instabilities (Toth & Kalnay, 1997). Practical EPSs have been developed for predicting weather forecasts (on the time-scale of days), El Niño events (seasonal prediction), and climate change projects (decadal predictions), although their evaluation and verification represent a major challenge (Atger, 1999). Since 1992, operational medium-range ensembles have been run by the European Centre for Medium-range Weather Forecasts and the National Centers for Environmental Prediction. EPSs have improved to the point where developments in weather prediction are now increasingly focused on severe or hazardous weather event prediction, such as heavy rainfall and flash floods (Buizza & Hollingsworth, 2000). However, there are a number of constraints that limit the seasonal forecast skill of dynamic models: errors affect numerical models of the ocean and atmosphere, ocean observations may be sparse, and uncertainty techniques remain in their infancy.

Emergency Flood Management in China

China's population and infrastructure are highly vulnerable to flood events: approximately half of the nation's population and three-quarters of its property are located in regions at risk for water-related disasters. Accordingly, flood DSSs are essential to protect the safety of people and property in China. China's worst flooding in over 40 years occurred in central China in 1998: monsoon rains affected up to 250 000 people and flooded over 20.2 million hectares (50 million acres), claiming more than 5000 lives. The 1998 flood significantly impacted the entire Yangtze River valley region. As shown in Figure 2, rainfall anomalies exceeded the climatological mean by more than 100%. In particular, high water levels lasted for more than 2 months, leading to one of the most costly and deadly flood disasters in history: the flood was responsible for over US$30 billion in economic losses and more than 3600 deaths.

The high flood levels forced Chinese officials to consider drastic strategies to save large cities on the Yangtze River from devastating flood events. During the summer of 1998 it was feared that Yangtze River flooding would cause a catastrophic dike failure in the densely populated city of Wuhan (a city of 7 million inhabitants in China's central Hubei province). To protect Wuhan and neighbouring rich farmland, the Chinese authorities deliberately destroyed dikes in Jianli County (Hubei province), about 145.2 km (90 miles) upriver from Wuhan. While the purposeful destruction of dikes at Jianli temporarily prevented Wuhan from being flooded, a difficult decision remained: whether or not to purposefully destroy dikes in the Yangtze River's Jingjiang section, thereby reducing the risk of the sudden bursting of Jingjiang River dikes at Shashi City. Accordingly, extensive preparations were put in place to dynamite the dikes and divert waters into the Jingjiang floodplain. This was expected to submerge more than 1000 km^2 (386 square miles) of land and 27 000–33 000 ha (68 000–82 000 acres) of farmland in the Jingjiang floodplain. Officials at the Jingjiang Flood Diversion Management Bureau awaited instructions from China's State Council, or Cabinet, on whether or not to destroy dikes and open floodgates. For this order to be given, water levels on the Yangtze would have to reach a record high of 45 m (149 feet) at the monitoring station in Shashi, just north of the area that would be flooded. Immediately after the destruction of dikes at Jianli, officials reported that the water level at Shashi fell slightly, alleviating the immediate need to open floodgates and destroy dikes in the crucial Jingjiang flood diversion area.

Many hydrologic models have been applied to the Yangtze River basin in order to improve flood modelling. Five models are now discussed, as described in Markar *et al.* (2002). The first two hydrologic models are non-linear rainfall run-off routing models: Urban Runoff Branching Structure (URBS) and Runoff Analysis and Flow Training Simulation (RAFTS). URBS is an Australian networked run-off routing model which estimates surface run-off and flood hydrographs by routing rainfall excess through a catchment storage module (Markar *et al.*, 2002). The total catchment hydrograph is obtained by adding baseflow to the surface run-off hydrograph. URBS has been used for

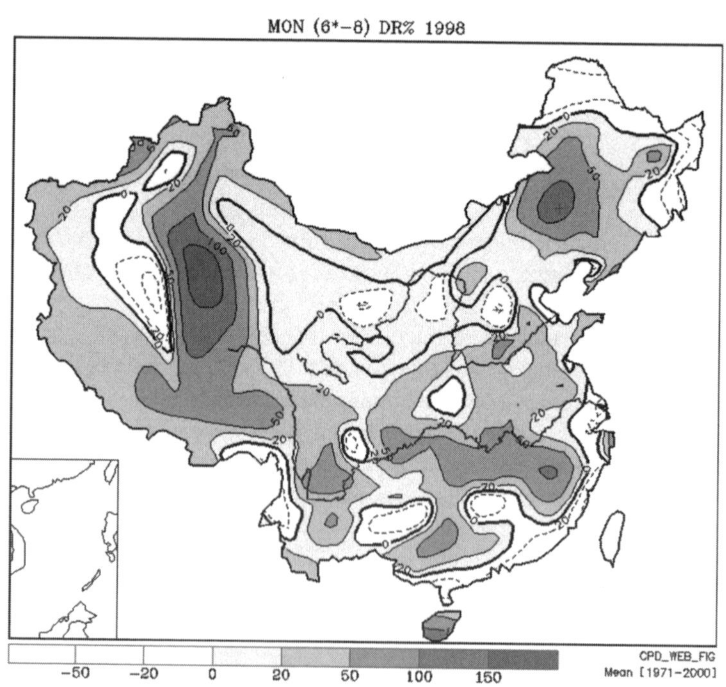

Figure 2. The observed percentage summer rainfall anomalies (averaged over June, July, and August) for 1998 in China.

over 15 years for real-time flood forecasting. It can be used as either an event or continuous model and adequately captures the distributed nature of catchment storage. RAFTS (formerly Regional Strom Water Model (RSWM)) is a networked rainfall run-off routing model that permits rainfall loss modelling and channel routing. Moreover, it estimates subcatchment flow by using Laurenson's non-linear run-off routing method. A third hydrologic model that has been tested in the Yangtze River basin is HEC-HMS (formerly HEC-1 flood hydrology model), developed at the HEC (US Army Corps of Engineers). It can be used as either an event or a continuous subcatchment model (rather than a lumped single node model). Here, excess rainfall is converted to direct run-off using unit hydrograph theory. The soil moisture accounting model is similar to that used in RAFTS, while the rainfall loss modelling approach is similar to URBS. Finally, HEC-HMS uses synthetic unit hydrograph parameters to represent catchment and storage characteristics (although the gridded modified Clark distributed model or kinematic wave model can also be used). Fourth, the Xinanjiang (XAJ) model was developed in China for hydrologic analysis of humid and semi-humid regions. Over the past quarter-century it has been extensively applied for flood forecasting and simulation using either synthetic or user defined unit hydrographs. One of its most valuable features is the use of a non-uniform spatial distribution to describe catchment soil moisture deficit and storage capacity. This allows for the simulation of run-off in terms of a non-uniform distribution. However, there are a number of limitations of the XAJ model: it is not able to incorporate land-use changes; and it is not possible to model flow diversions. Finally, XAJ can only be used as a lumped single node catchment model. A fifth hydrologic model, the Antecedent Precipitation Index (API), is well suited for flood forecasting in the Yangtze River valley and other catchments in China. API estimates direct run-off using empirical correlation flood forecasting methods. A linear reservoir approach is used to separately calculate baseflow, which is then added to the surface run-off hydrograph to determine the total run-off hydrograph.

The flood control decision support system (FCDSS) constitutes an important and timely flood management DSS (Siyi *et al.*, 1996). Zhou *et al.* (2004) discuss FCDSSs in the context of web-based GIS for emergency flood control systems (EFCSs) with particular emphasis on emergency management and information retrieval in China. Specifically, EFCSs consists of six important modules: database management; emergency information query; site location; emergency material operation; 'earth–rock' volume calculations; and three-dimensional scene visualization.

Emergency levee repair is a key component of flood risk management in China: in the event of a flood, levees may need to be strengthened and repaired. In addition, temporary levees may be required. To facilitate the construction of levees, it is important to determine the volume of earth and rock required for emergency levee repair, the type and number of vehicles required to transport construction material, and the optimal process for storing and transporting earth and rock to the exact flood location. If the local volume of earth and rock is insufficient to meet the requirements of emergency levee repair, then the flood manager may require information about the location of material warehouses and the nature of the road network (optimal routes, road network quality, etc.). Specifically, EFCSs are capable of automatically determining elevation data of the ground surface where the levee is being repaired by means of a digital elevation model (DEM).

In addition, three-dimensional scene visualization is valuable in helping planners and managers improve flood risk management. Zhou *et al.* (2004) use a five-step approach to

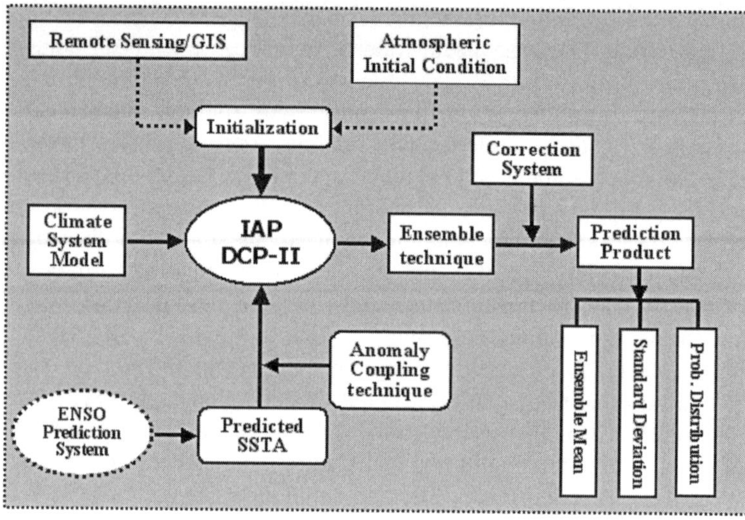

Figure 3. Configuration of IAP DCP-II. *Note:* SSTA, Sea surface temperature anomaly.

create realistic three-dimensional scene visualization in the context of levee construction in China. First, terrain vector data are downloaded from the spatial database. Second, data downloaded by the spatial server are transferred into DEM data by interpolation techniques. Third, a geographic co-ordinate system and a projected co-ordinate system are created in order to further process DEM data. Fourth, a three-dimensional model of the levee system under consideration is developed. Fifth, three-dimensional scene visualization simulation is achieved with a three-dimensional viewing program (Xiao *et al.*, 2001). A rich graphical user interface can help flood decision makers to perform a number of valuable flood analyses, including a determination of detailed emergency operation plans.

Finally, a dynamic seasonal climate prediction system from the Institute of Atmospheric Physics, Chinese Academy of Sciences (IAP), has been successfully used to improve flood prevention throughout China, including the Yangtze River valley. In 1989, the IAP dynamic seasonal prediction system was first applied to seasonal and extra-seasonal predictions of the summer monsoon rainfall anomaly with use of an AGCM. The second generation IAP dynamic climate prediction system (IAP DCP-II) consists of five components, as shown in Figure 3: (1) an ENSO prediction system (El Niño constitutes an important source of potentially predictable inter-annual rainfall variability); (b) the prediction of integrations and an anomaly coupling technique; (3) an ensemble prediction technique; (4) a correction system; and (5) prediction products and analyses. Experiments have shown that the IAP system is capable of predicting large positive and negative anomalies in summer rainfall, such as the severe flooding in the Huaihe River and Yangtze River regions in 1991.

Conclusions

The proposed DSS architecture is a valuable non-structural approach to improve flood planning and management. The latest advances in decision support, remote sensing/GIS, MCDM models, and seasonal climate prediction were discussed in order to reduce the vulnerability of society to floods. In particular, flood DSSs can facilitate communication and knowledge transfer, improve forecast skill, and increase the transparency of strategic flood decision management, making the links among knowledge, assumptions, and choices more explicit. The devastating 1998 Yangtze flood illustrates the complexity and importance of flood events (including large-scale evacuation and the deliberate destruction of dikes). In the context of this flood, numerical weather/climate models for predicting severe rainfall anomalies (from several days to several months in advance) were described. Specifically, the IAP DCP-II employs ensemble prediction with dynamically conditioned perturbations to reduce the uncertainty associated with seasonal climate prediction. Such predictions are extremely valuable for flood DSSs, allowing time for proactive flood protection measures to be taken. In conclusion, integrated flood DSSs can help to improve flood risk analysis under increasingly uncertain hydro-meteorological variability and change.

References

Al-Sabhan, W., Mulligan, M. & Blackburn, G. A. (2003) A real-time hydrological model for flood prediction using GIS and the WWW, *Computer, Environment, and Urban Systems*, 27, pp. 9–32.

American Water Resources Association (2005) Letter to the President of the United States, in: G. Galloway (Ed.) *American Water Resources Association's Second National Water Resources Policy Dialogue* (Tucson: American Water Resources Association).

Annan, K. (2005) Address of the UN Secretary-General Kofi Annan to the high-level segment of the International Meeting for the 10-year Review, available at http://un.by/en/news/world/2005/12-01-05-02.html, accessed August 2005.

Appleton, B. (2002) *Climate Changes the Water Rules: Dialogue on Water and Climate* (Liverpool: Printfine).

Atger, F. (1999) The skill of ensemble prediction systems, *Monthly Weather Review*, 127, pp. 1941–1953.

Barnston, A. G. (1994) Linear statistical short-term climate predictive skill in the Northern Hemisphere, *Journal of Climate*, 7, pp. 1513–1564.

Barnston, A. G. & Ropelewski, C. F. (1992) Prediction of ENSO episodes using canonical correlation analysis, *Journal of Climate*, 5, pp. 1316–1345.

Barton, I. & Bathols, J. (1989) Monitoring floods with AVHRR, *International Journal of Remote Sensing*, 10, pp. 1873–1892.

Bengtsson, L., Schlese, U., Roeckner, E., Latif, M., Barnett, T. P. & Graham, N. (1993) A two-tiered approach to long-range climate forecasting, *Science*, 261, pp. 1026–1029.

Buizza, R. & Hollingsworth, A. (2000) severe weather prediction using the ECMWF EPS, *ECMWF Newsletter*, 89, pp. 2–12.

Charney, J. G. & Shukla, J. (1981) *Monsoon Dynamics* (New York: Cambridge University Press).

Clark, M. J. (1998) Putting water in its place: a perspective on GIS in hydrology and water management, *Hydrological Processes*, 12, pp. 823–834.

Consortium of Universities for the Advancement of Hydrologic Science (2004) Available at http://cuahsi.sdsc.edu/html/index.html, accessed August 2005.

Correia, F. N., Saraiva, M. D., Da Silva, F. N. & Ramos, I. (1999) Floodplain management in urban developing areas: GIS-based flood analysis and urban growth modeling, *Water Resources Management*, 13, pp. 23–37.

Deutsch, M., Ruggles, F., Guss, P. & Yost, E. (1973) Mapping the 1973 Mississippi floods from the Earth Resource Technology satellites, *Proceedings of International Symposium on Remote Sensing and Water Resource Management* (Burlington, VT: American Water Resources Association).

Folland, C. K., Owen, J. A., Ward, M. N. & Colman, A. W. (1991) Prediction of seasonal rainfall in the Sahel region of Africa using empirical and dynamic methods, *Journal of Forecasting*, 10, pp. 2–56.

Frank, A. U. (1993) The use of geographical information system: the user interface is the system, in: D. Medyckyj-Scott & H. Hearnshaw (Eds) *Human Factors in Geographical Information Systems*, Chapter 18 (London: Belhaven Press).

Hajat, S., Erbi, K. L., Kovats, S., Menne, B., Edwards, S. & Haines, A. (2003) The human health consequences of flooding in Europe and the implications for public health, *Journal of Applied Environmental Science and Public Health*, 1, pp. 13–21.

Hall, J. W., Meadowcroft, I. C., Sayers, P. B. & Bramley, M. E. (2003) Integrated flood risk management in England and Wales, *Natural Hazards Review*, 4, pp. 126–135.

Hammer, G. L., Holzworth, D. P. & Stone, R. C. (1996) The value of skill in seasonal climate forecasting to wheat crop management in a region with high climatic variability, *Australian Journal of Agricultural Research*, 47, pp. 717–737.

Hastenrath, S. (1995) Recent advances in tropical climate prediction, *Journal of Climate*, 8, pp. 1519–1532.

Intergovernmental Panel on Climate Change (2001) *Climate Change 2001: The Scientific Basis* (Cambridge: Cambridge University Press).

Islam, M. M. & Sadu, K. (2002) Development of priority map remote sensing data for flood counter measures by geographic information system, *Journal of Hydrological Engineering*, 7, pp. 346–355.

Ji, M., Kumar, A. & Leetmaa, A. (1994) A multi-seasonal climate forecast system at the National Meteorology Centre, *Bulletin of the American Meteorological Society*, 75, pp. 569–577.

Jonkman, S. N. (2005) Global perspectives on loss of human life caused by floods, *Natural Hazards*, 34, pp. 151–175.

Kingston, S., Carver, S., Evans, A. & Turton, I. (2000) Web-based public participation geographical information systems: an aid to local environment decision-making, *Computer, Environment and Urban Systems*, 24, pp. 109–122.

Kite, G. W. (1988) *Frequency and Risk Analysis in Hydrology* (Littleton, CO: Water Resources Publications).

Knaff, J. A. & Landsea, C. W. (1997) An El Niño Southern Oscillation climatology and persistence (CLIPER) forecasting scheme, *Weather Forecasting*, 12, pp. 633–652.

Krishna, K., Kleeman, R., Cane, M. A. & Rajagopalan, B. (1999) Epochal changes in Indian monsoon–ENSO precursors, *Geophysical Research Letters*, 26, pp. 75–78.

Landman, W. A., Mason, S. J., Tyson, P. D. & Tennant, W. J. (2001) Retro-active skill of multi-tiered forecasts of summer rainfall over southern Africa, *International Journal of Climatology*, 21, pp. 1–19.

Lin, Z., Sun, J., Wei, J., Zhou, G., Zhang, X., Chen, H. & Lang, X. (2003) Real-time weather and seasonal climate predictions for 2002 summer and their verifications, *Climatic and Environmental Research*, 8, pp. 241–257.

Livezey, R. E. & Barnston, A. G. (1988) An operational multifield analog prediction system for the United States seasonal temperature, part I: system design and winter experiments, *Journal of Geophysical Research*, 93, pp. 19053–19074.

Lorenz, E. N. (1969) The predictability of a flow which possesses many scales of motion, *Tellus*, 21, pp. 289–307.

Markar, M. S., Kwan, R. T. F., Conroy, A. L., Mingxin, L., Jing, Z. & Lina, A. (2002) Evaluation of five hydrologic models for real-time flood forecasting use in the Yangtze River catchment. *Proceedings of the Second International Symposium on Flood Defence*, Beijing, China, 10–13 September 2002, pp. 901–907.

Mason, S. J., Goddard, L., Graham, N. E., Yulaeva, E., Sun, L. & Arkin, P. A. (1999) The IRI seasonal climate prediction system and the 1997/98 El Niño event, *Bulletin of the American Meteorological Society*, 80, pp. 1853–1873.

Meehl, G. A., Washington, W. M., Collins, W. D., Arblaster, J. M., Hu, A., Buja, L. E., Strand, W. G. & Teng, H. (2005) How much more global warming and sea level rise?, *Science*, 307, pp. 1769–1772.

Mileti, D. (1999) *Disasters by Design: A Reassessment of Natural Hazards in the United States* (Washington, DC: Joseph Henry Press).

Miller, R. C., Guertin, D. P. & Heilman, P. (2004) Information technology in watershed management decision making, *Journal of the American Water Resources Association*, 40, pp. 347–358.

Molteni, F., Buizza, R., Palmer, T. N. & Petroliagis, T. (1996) The ECMWF ensemble prediction system: methodology and validation, *Quarterly Journal of the Royal Meteorological Society*, 122, pp. 73–119.

Murray, D. (2005) Earth Policy Institute resources on ice melting. Available at http://www.earth-policy.org/Idicates/Ice/index.htm, accessed August 2005.

Myers, M. F. & White, G. F. (1993) The challenge of the Mississippi flood, *Environment*, 35, pp. 6–35.

Mysiak, J., Giupponib, C. & Rosatoc, P. (2005) Towards the development of a decision support system for water resource management, *Environmental Modelling and Software*, 20, pp. 203–214.

National Assessment Synthesis Team (2000) *Change Impacts on the United States: The Potential Consequences of Climate Variability and Change* (New York: Cambridge University Press).

National Audit Office (2001) *Inland Flood Defence: Report by the Comptroller and Auditor General of the National Audit Office* (London: Stationery Office).

National Research Council (1982) *Levee Policy for the National Flood Insurance Policy* (Washington, DC: National Academies Press).

Nilsson, C., Reidy, C. A., Dynesius, M. & Revenga, C. (2005) Fragmentation and flow regulation of the world's large river systems, *Science*, 308, pp. 405–408.

Ott, T. & Swiaczny, F. (2001) *Time-integrative Geographic Information Systems* (Berlin: Springer).

Palmer, T. N. & Anderson, D. L. T. (1994) The prospects for seasonal forecasting, *Quarterly Journal of the Royal Meteorological Society*, 120, pp. 755–793.

Palmer, T. N., Brankovic, C. & Richardson, D. S. (2000) A probability and decision-model analysis of PROVOST seasonal multi-model ensemble integrations, *Quarterly Journal of the Royal Meteorological Society*, 126, pp. 2013–2034.

Pinter, N. (2005) One step forward, two steps back on US floodplains, *Science*, 308(5719), pp. 207–208.

Quayle, E. T. (1929) Long-range rainfall forecasting from tropical (Darwin) air pressure, *Proceedings of the Royal Society*, 41, pp. 160–164.

Rango, A. & Anderson, A. T. (1974) Flood hazard studies in the Mississippi River basin using remote sensing, *Water Resource Bulletin*, 10, pp. 1060–1081.

Rosati, A., Miyakoda, K. & Gudgel, R. (1997) The impact of ocean initial conditions on ENSO forecasting with a coupled model, *Monthly Weather Review*, 125, pp. 754–772.

Saaty, T. L. (2004) Decision making—the analytic hierarchy and network processes (AHP/ANP), *Journal of Systems Science and Systems Engineering*, 13, pp. 1–34.

Sahai, A. K., Soman, M. K. & Satyan, V. (2000) All India summer monsoon rainfall prediction using an artificial neural network, *Climate Dynamics*, 16, pp. 291–302.

Sanyal, J. & Lu, X. (2004) Application of remote sensing in flood management with special reference to monsoon Asia: a review, *Natural Hazards*, 33, pp. 283–301.

Simonovic, S. P. & Ahmad, S. (2005) Computer-based model for flood evacuation emergency planning, *Natural Hazards*, 34, pp. 25–51.

Siyi, H., Dedun, S. & Wy, Y. (1996) Framework and design of the Yangtze River flood control decision support system, *Advances in Water Science*, 7, pp. 283–294.

Smith, L. C. (1997) Satellite remote sensing of river inundation area, stage and discharge: a review, *Hydrological Processes*, 11, pp. 1427–1439.

Stedinger, J. R., Vogel, R. M. & Foufoula-Georgiou, E. (1993) Frequency analysis of extreme events, in: D. R. Maidment (Ed.) *Handbook of Hydrology* (New York: McGraw-Hill).

Stockdale, T. N. (2000) An overview of techniques for seasonal forecasting, *Stochastic Environmental Research and Risk Assessment*, 14, pp. 305–318.

Stone, R. C., Hammer, G. L. & Marcussen, T. (1996) Prediction of global rainfall probabilities using phases of the Southern Oscillation Index, *Nature*, 384, pp. 252–255.

Tangang, F. T., Tang, B., Monahan, A. H. & Hsieh, W. W. (1998) Forecasting ENSO events: a neural network—extended EOF approach, *Journal of Climate*, 11, pp. 24–41.

Tholey, N., Clandillon, S. & De Fraipont, P. (1997) The contribution of space borne SAR and optical data in monitoring flood events: examples in northern and southern France, *Hydrological Processes*, 11, pp. 1427–1439.

Toth, Z. & Kalnay, E. (1997) Ensemble forecasting at NCEP and the breeding method, *Monthly Weather Review*, 125, pp. 3297–3319.

Vermieren, J. C. & Watson, C. C. (2001) New technology for improved storm risk assessment in the Caribbean, *Disaster Management*, 6, pp. 191–196.

Wang, Y., Colby, J. D. & Mulcahy, K. A. (2002) An efficient method for mapping flood extent in a coastal flood plain using Landsat TM and DEM data, *International Journal of Remote Sensing*, 23, pp. 3681–3696.

Ward, N. M. & Folland, C. K. (1991) Prediction of seasonal rainfall in the north Nordeste of Brazil using eigenvectors of sea-surface temperatures, *International Journal of Climatology*, 11, pp. 711–743.

Wei, Y. M., Fan, Y., Lu, C. & Tsai, H. T. (2004) The assessment of vulnerability to natural disasters in China by using the DEA method, *Environmental Impact Assessment Review*, 24, pp. 427–439.

Werstedt, K. & Hersh, R. (2002) Climate forecasts in flood planning: promise and ambiguity, *Journal of the American Water Resources Association*, 38, pp. 1703–1713.

Wigley, T. M. L. (2005) The climate change commitment, *Science*, 307, pp. 1766–1769.

Wu, R. & Wang, B. (2002) A contrast of the East Asia summer monsoon–ENSO relationship between 1962–1977 and 1978–1993, *Journal of Climate*, 15, pp. 3266–3279.

Xiao, L., Zhong, E., Liu, J. & Soung, G. (2001) A discussion on basic problems of 3D GIS, *Journal of Image and Graphics*, 6, pp. 843–848.

Zerger, A. & Wealands, S. (2004) Beyond modelling: linking models with GIS for flood risk management, *Natural Hazards*, 33, pp. 191–208.

Zerger, A., Smith, D. I., Hunter, G. J. & Jones, S. D. (2002) Riding the storm: a comparison of uncertainty modeling techniques for storm surge risk management, *Applied Geography*, 22, pp. 307–330.

Zhang, X. G. & Casey, T. M. (1992) Long-term variations in the Southern Oscillation and relationships with Australian rainfall, *Australian Meteorological Magazine*, 40, pp. 211–225.

Zhou, H., Bin, H., Zhu, L. & Liang, G. (2004) Application of web GIS in emergency flood control system, *Proceedings of the International Conference of GIS and Remote Sensing in Hydrology, Water Resources, and Environment*, Three Gorges Dam, China, International Association of Hydrological Sciences (IAHS) Publication 289.

Integrated Risk Management of Flood Disasters in Metropolitan Areas of China

PEIJUN SHI, YI GE, YI YUAN & WEIPING GUO

Introduction

Large-scale human activities have significantly changed the hydrological cycle in natural conditions and, to some extent, have made a great change in the stability of this process, and also aggravated the problems of regional water shortage, flood disaster, and water pollution. In land use and land cover change research, scientists have paid more attention to how land use and land cover changes affect water circulation, water supply, and water quality (Lambin *et al.*, 1999). The urbanization process, especially landscape urbanization in metropolitan areas, has directly changed regional landscape structures on a large scale and caused a sharp decrease in the ratio of wetlands and water bodies. As a result, the natural water level is overlapped by the 'social water level',[1] and useful precipitation cannot be conserved. It turns into a flood disaster and further heightens the flood vulnerability of the metropolis (Shi *et al.*, 2002a). There are many relevant research studies on floods and urbanization. The report of the United Nations Educational, Scientific, and Cultural Organization (1974) systematically explained the hydrological effects in the process of urbanization, among which the most crucial effects are seepage reduction and peak-flow increase. Hall's (1984) research pointed out that the main hydrological responses to the urbanization process are the problems of the river head, flood control, and sewage management. Brun & Band (2000) also discussed the impact of urbanization on run-off yield and concentration. They particularly emphasized that the increased run-off coefficient accelerates the flow concentration process, heightens flood peak, and shortens concentration time. All of these show that urbanization has changed local run-off and directly affected the formation of urban floods. In metropolitan areas, the urbanization process as landscape change actually affects the hydrological process and precipitation change (Shi *et al.*, 2000; Fu, 2001).

Drawing on the discussion of how to balance flood disaster magnitude and vulnerability in metropolitan areas (Shi *et al.*, 2002a), this paper attempts to determine the relationship between landscape urbanization and flood disaster changes using two case studies: run-off yield and concentration in Shenzhen; and water-level changes in urban water bodies (including wetlands and low-lying lands) in Shanghai. Then, the paper puts forward a solution to alleviate the increase in fluctuating flood disaster in metropolitan areas, through adjustments in land use structure and pattern. Finally, the authors propose an optimal approach for integrated risk management of flood disaster in Chinese metropolitan areas.

Urbanization Process and Water Problems in China

The Relationship between Water Resource Shortage and Flood Disasters

Water is a crucial resource for sustainable development. In recent years it has become one of the scarce resources limiting local development in China. The gross amount of water in China is large, but the per capita share is small. Thus, there is a distinct conflict between supply and demand (Liu & He, 1996). In particular, in the urbanization process, the amount of water supplied by water conservancy facilities cannot meet the growing demand for agriculture, industry, and residential use (Shi, 1997). Since the 1990s, more than 300 cities in China have become short of water, of which 144 cities are in an extremely serious situation. For example, in the northern metropolitan areas, the water shortage in the central Liaoning area is 1.5 billion m^3; in the Beijing–Tianjin–Tangshan area in a normal year it is 2.0 billion m^3, and in a dry year it is 4.0 billion m^3; and in the Shandong Peninsula it is 2.2 billion m^3 in a normal year, while in a dry year it reaches 2.6 billion m^3 (Shi, 1997).

Lack of water resources has increased the risk of local flood disaster greatly. Overexploitation of groundwater triggered by water shortage has resulted in serious land subsidence. Now, there are different funnel areas of groundwater in the North China plain, the Huang-Huai-Hai plain, and the Yangtze delta area. They have extended to almost 90 000 km^2, accounting for 70% of the total plain areas. Among these, a water-bearing layer of 10 000 km^2 has been exhausted and has formed a deep groundwater funnel area with a separate centre of Tianjin, Hengshui, Cangzhou, and Langfang and with a total area of 56 000 km^2 (Liu, 2004). Surface elevation loss caused by subsidence has surpassed the impact of sea-level change 10-fold (Zhang & Shi, 1994). Thus, the problem of the so-called 'little flood, serious consequence' appears and the risk of local flood disaster increases greatly. On the other hand, in metropolitan areas, floods are not often caused by absolute surplus water alone, but are also due to the irrational land use structure and pattern, which turn useful precipitation into flood. Hence, if we can adjust the land use structure and pattern, relieving water shortage and flood disasters, a win–win situation will result.

The Intensification of Flood Disaster in China

With the development of urbanization, flood disaster in metropolitan areas has been intensifying and further threatening regional sustainable development.

The Increase in the Incidence of Floods

Flood frequency has obviously increased with rapid urbanization since 1978. From 1949 to 1966, the flood frequency of each county was less than 50 times/100 years, and only 25 counties reached 20–50 times/100 years; however, from 1976 to 1998, the flood frequency of one county has exceeded 50 times/100 years, and 125 counties have reached 20–50 times/100 years (Figure 1).

As shown in Figure 1, the 10 metropolitan areas are mainly located in the middle or lower reaches of major rivers and are thus more vulnerable to flood disaster. Since reforming and opening up in 1978, except in the Jing–Jin–Tang area and the central China area, flood incidences in the remaining eight metropolitan areas have increased markedly (Figure 2). Comparing this result with annual precipitation in the same areas (Figure 3), it is found that precipitation did not affect the incidence of floods.

The Increase in Flood Losses

Since 1978, though the capacity of flood control in China has been strengthened considerably, flood disaster losses have not decreased correspondingly. There was no large flood in the drainage basins, but flood-affected areas and flood damage percentages have continued to increase (Table 1). Moreover, the annual direct economic losses from floods in 1991 reached almost US$10 billion, while the one in 1998 resulted in US$30 billion more, a dramatic increase.

Cases of Flood Disaster Intensification in Metropolises

The Case of Shenzhen

In the Shenzhen Region there are 160 rivers, of which 13 have drainage basin areas larger than 10 km^2 and five larger than 100 km^2. The authors selected the Buji River basin as their study area. The Buji River flows through the town of Buji to the downtown area of Shenzhen City. The length of the Buji River is 17.7 km and the total area of its drainage basin is 57 km^2. In recent years, with economic development, the urbanization level in the Buji River basin has been rising rapidly.

Data and Methods

Relevant land use/cover data (Table 2) were acquired from Landsat images in 1980 (Landsat's Multi Spectral Scanner), 1988 (Landsat's Thematic Mapper (TM)), 1994 (TM), and 2000 (TM). Meteorological data including annual and monthly precipitation

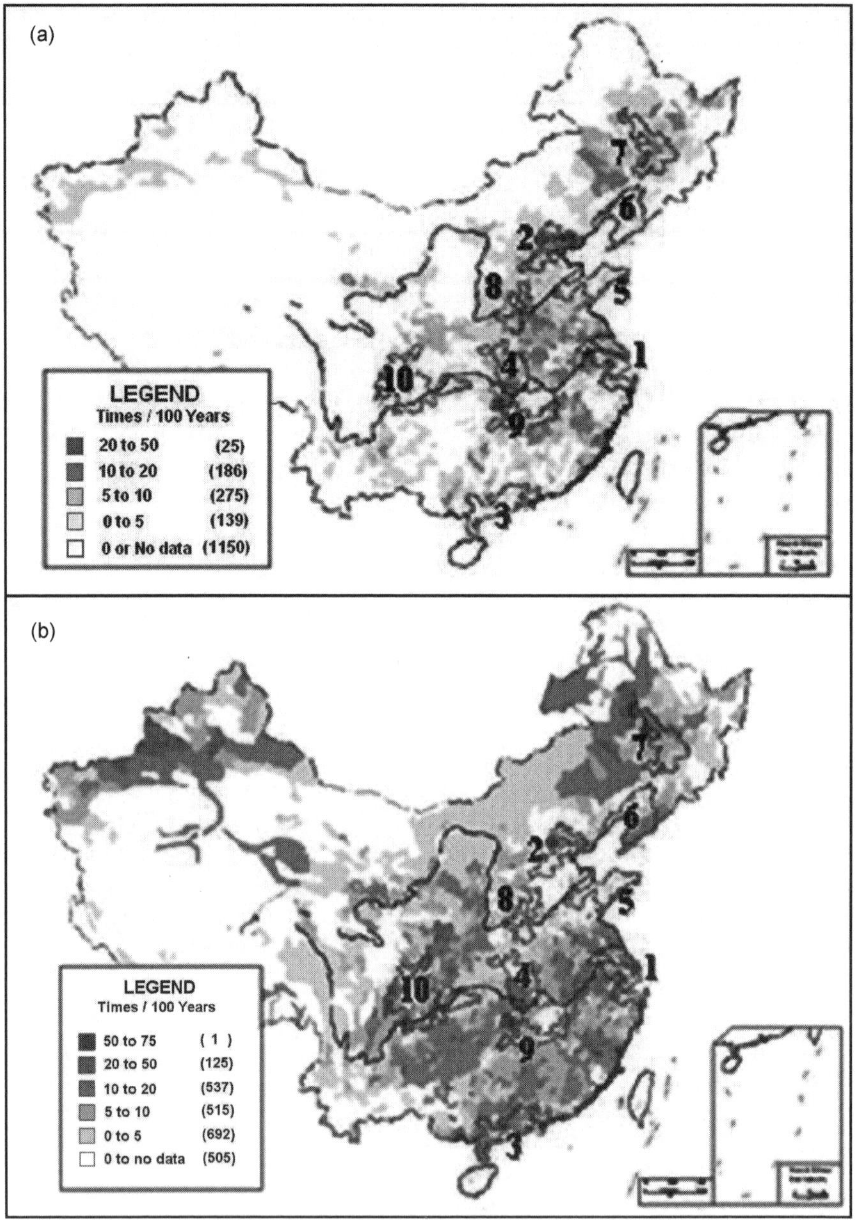

Figure 1. Incidence of county floods in China: a, 1949–66; b, 1976–98.

(1953–2000), average precipitation and run-off, annual maximum rainfall in 24 h (1954–93), and precipitation data for some typical rainstorms were obtained from the local meteorological stations. Moreover, social data from Shenzhen statistical yearbooks and landscape urbanization data collected by Global Position System (GPS) were also used.

As the hydrological and meteorological data are not extensive enough in our research, it is unsuitable to simulate the process of run-off yield and concentration through common

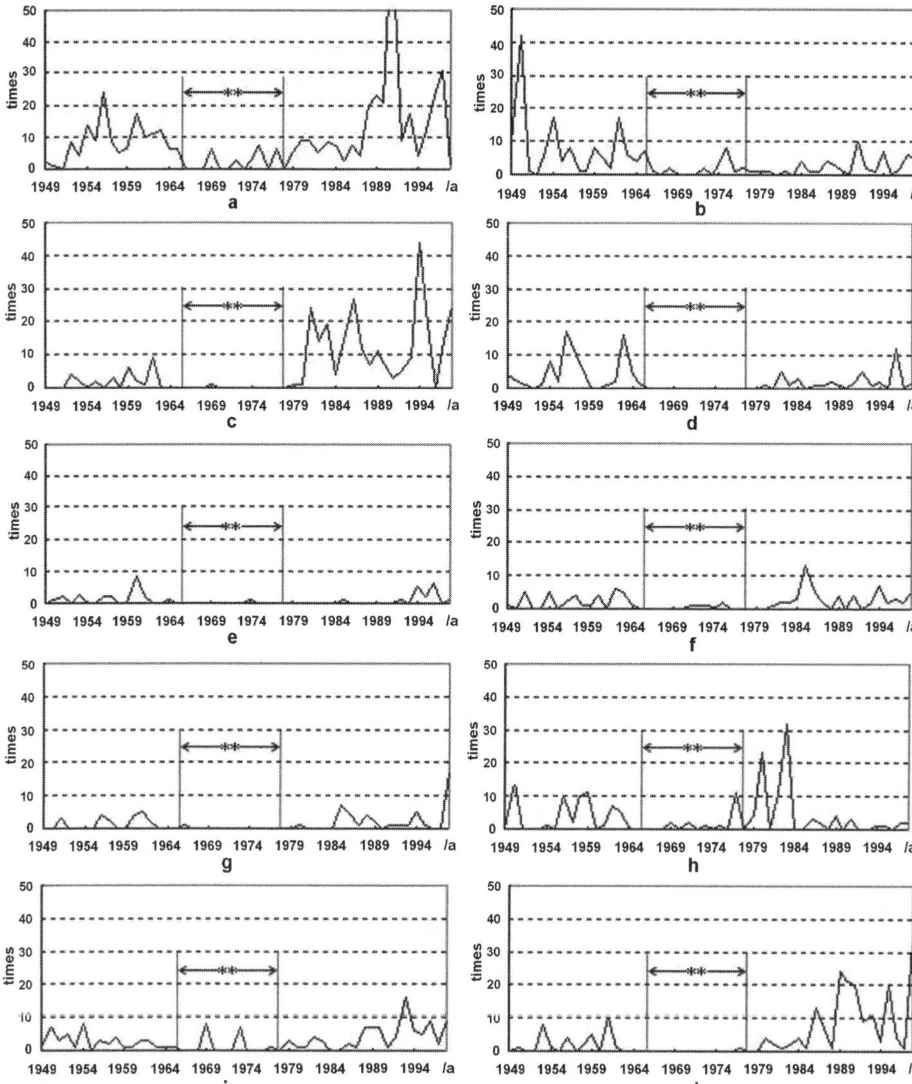

Figure 2. Incidence of floods in each metropolitan area in China: a, Yangtze delta; b, Beijing–Tianjin–Tangshan; c, Pearl River delta; d, central China; e, Shandong Peninsula; f, central Liaoning; g, Harbin–Daqing–Qiqihar; h, Wuhan; i, central Hunan; j, Sichuan basin. **Unreliable data during Cultural Revolution. *Source:* Natural Disaster Database of China, established by Key Laboratory of Environment Change and Natural Disaster, Ministry of Education, Beijing Normal University.

hydrological methods. Furthermore, the parameters and structure of traditional hydrological models are not adaptable to data from remote sensing (Wang, 1989; Lu, 1990) because few of them take land use condition as an influencing factor directly or indirectly. Therefore, it is advisable to use a hydrological model based on remote sensing to simulate the run-off process in the Buji River basin. The US Department of Agriculture Soil Conservation Service (SCS) model (SCS, 1972) has been widely used for water

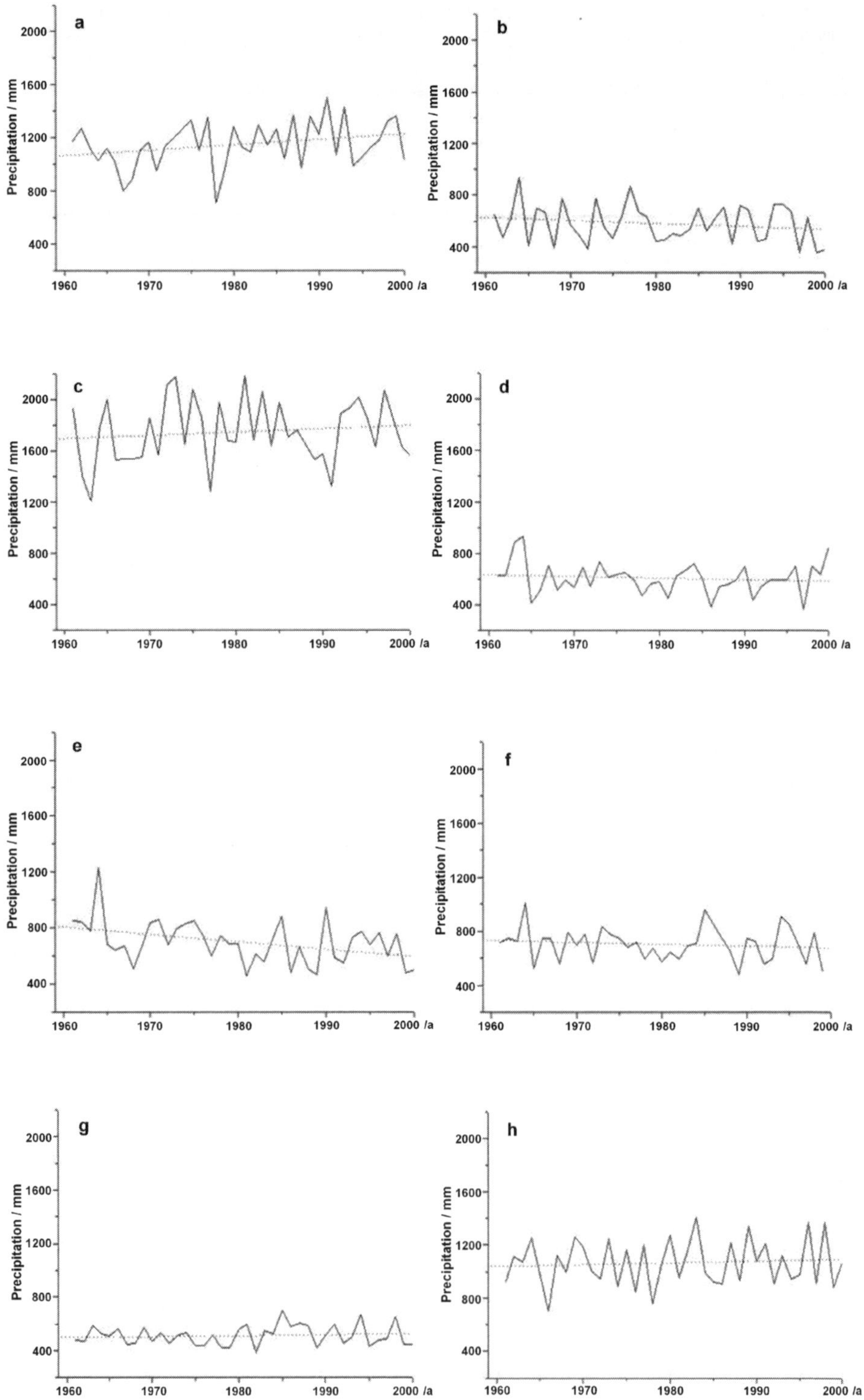
a
b
c
d
e
f
g
h
Precipitation / mm
400
800
1200
1600
2000
1960
1970
1980
1990
2000 /a

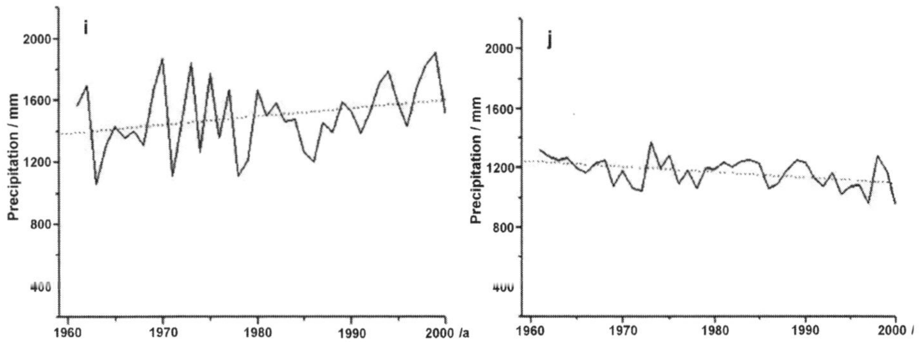

Figure 3. Precipitation in each metropolitan area in China: a, Yangtze delta; b, Beijing–Tianjin–Tangshan; c, Pearl River delta; d, central China; e, Shandong Peninsula; f, central Liaoning; g, Harbin–Daqing–Qiqihar; h, Wuhan; i, central Hunan; j, Sichuan basin. *Source:* Natural Disaster Database of China, established by Key Laboratory of Environment Change and Natural Disaster, Ministry of Education, Beijing Normal University.

resources management and planning. It has also been introduced and applied in China. Since the major input parameters for the SCS model are land use and soil type, the SCS model is potentially compatible with remote-sensing input (Lewis *et al.*, 2000). The SCS model can be expressed as:

$$\left.\begin{array}{ll} Q = \frac{(P-0.2S)^2}{P+0.8S}, & P \geq 0.2S \\ Q = 0, & P < 0.2S \end{array}\right\} \tag{1}$$

where Q is direct run-off, P is storm rainfall, and S is potential maximum retention or infiltration. For convenience and standardization, S is expressed in the form of a dimensionless run-off curve number (CN):

$$S = \frac{25400}{CN} - 254 \tag{2}$$

where CN represents the run-off potential of the land cover–soil complex, governed by soil property, cover type, and the hydrological condition of the land surface. Hence, it can reflect the impact of human activities on run-off indirectly, and link the hydrological model with remote-sensing applications. CN also depends on the antecedent wetness of the drainage basin. Three antecedent moisture conditions (AMCs) were defined as dry, moderate, and wet, and denoted as AMCI, AMCII, and AMCIII, respectively. The soil

Table 1. Flood disaster damage in China (1950–89)

	Flood-affected		Flood-damaged		
Period	Area (km^2)	Percentage increase	Area (km^2)	Damage ratio (%)	Percentage increase
1950–66	80 000	0	50 000	62.5	0
1970–78	52 000	− 35	22 000	42.3	− 20.2
1979–89	100 000	92.3	52 600	52.6	10.3

Note: In 1954, a large flood of the entire Yangtze River drainage occurred.
Source: Liu & He (1996).

Table 2. Land use types in the Buji River basin: 1980, 1988, 1994, and 2000

Year	Index[a]	Urban land with high density	Urban land with middle and low density	Farmland	Orchard	Forest	Grassland	Water	Wetland	Unused land
1980	*A* (km^2)	0.00	2.02	18.50	0.00	24.80	7.64	1.15	0.31	2.77
	P (%)	0.00	3.53	32.35	0.00	43.37	13.36	2.01	0.54	4.84
1988	*A* (km^2)	1.28	20.99	4.45	8.77	17.89	0.74	0.56	0.01	2.48
	P (%)	2.23	36.71	7.79	15.33	31.29	1.30	0.99	0.02	4.34
1994	*A* (km^2)	2.85	25.04	2.27	6.59	13.73	0.00	0.70	0.02	5.98
	P (%)	4.98	43.80	3.97	11.53	24.01	0.00	1.22	0.04	10.46
2000	*A* (km^2)	6.08	27.50	1.52	4.45	13.51	0.00	0.87	0.02	3.24
	P (%)	10.63	48.09	2.65	7.77	23.63	0.00	1.52	0.04	5.66

[a] *A* represents the area of land; *P* represents the proportion of the total area of the area of a specific land use.

data were compiled from Guangzhou Institute of Geography Science (1983), according to soil classification categories in the SCS model. Based on this information, CN values in the Buji River basin were estimated and are shown in Table 3.

Results and Analyses

Impact of Landscape Urbanization on Run-off Yield in the Buji River Basin

The run-off coefficient is an index reflecting hydrological characteristics on different land surfaces. In this study, the run-off coefficient in each pixel was calculated using the SCS model. Run-off coefficients for designed storm probabilities of 90%, 50%, and 10%, respectively, were obtained (Table 4). From 1980 to 2000, under the dry antecedent soil condition, the run-off coefficient increased by 33.49%, 20.65%, and 12.57% for the 90%,

Table 3. Estimated CN values in the Buji River basin

	Soil and hydrology classification of CN											
	A			B			C			D		
Land use type	I	II	III	I	II	III	I	II	III	I	II	III
Urban land of high density	79	90	97	84	93	98	86	94	98	88	95	99
Urban land of middle or low density	39	60	80	56	74	90	68	83	94	74	87	96
Farmland	47	67	85	61	78	92	71	85	95	78	89	97
Garden plot	22	40	61	42	62	81	58	76	91	67	82	94
Forest	13	25	44	35	55	76	51	70	87	59	77	91
Bush and grass	20	36	57	39	60	80	56	74	90	64	80	93
Wetland	17	32	52	38	58	78	53	72	88	62	79	92
Unoccupied land	53	72	88	67	82	94	76	88	96	79	90	97
Water	94	98	100	94	98	100	94	98	100	94	98	100

Note: 1. In this Table, I, II, III respectively represent three early humid phases of soil: AMCI (relatively dry), AMC II (moderate) and AMC III (relatively wet). 2. Land surface infiltration categories: A, high infiltration rate; B, moderate infiltration rate; C, low infiltration rate; D, no infiltration.

Table 4. Integrated run-off coefficients of designed storm run-offs in the Buji River basin

		Antecedent soil moisture conditions			
Rainfall frequency	Year	AMCI	AMCII	AMCIII	AMCI → AMCIII
90%	1980	0.215	0.457	0.742	71.02%
	1988	0.227	0.477	0.76	70.13%
	1994	0.267	0.517	0.786	66.03%
	2000	0.287	0.533	0.793	63.81%
	1980 → 2000	33.49%	16.63%	6.87%	—
50%	1980	0.368	0.609	0.832	55.77%
	1988	0.387	0.628	0.845	54.20%
	1994	0.427	0.66	0.863	50.52%
	2000	0.444	0.672	0.868	48.85%
	1980 → 2000	20.65%	10.34%	4.33%	—
10%	1980	0.541	0.743	0.899	39.82%
	1988	0.56	0.759	0.907	38.26%
	1994	0.596	0.782	0.918	35.08%
	2000	0.609	0.789	0.921	33.88%
	1980 → 2000	12.57%	6.19%	2.45%	—
90% → 10%	1980	151.63%	62.58%	21.16%	—
	1988	146.70%	59.12%	19.34%	—
	1994	123.22%	51.26%	16.79%	—
	2000	112.20%	48.03%	16.14%	—

Note: 1980 → 2000 represents the comparative change of runoff coefficient when the condition of land use in 1980 changed into that in 2000; AMCI: low soil moisture; AMCII: moderate soil moisture; AMCIII: high soil moisture.

50%, and 10% storm probabilities, respectively; under the wet antecedent soil condition, it increased by 6.87%, 4.33%, and 2.45%, respectively. Therefore, urbanization could lead to higher run-off coefficients and thus to a greater risk of flooding.

Impact of Landscape Urbanization on Water Concentration in the Buji River Basin

First, rainfall intensities of 1%, 2%, and 5% were simulated and the results validated according to the designed floods reported in Shenzhen Water Resources Bureau (1994). Then, according to the SCS model, 24 h flood peak discharge in outlet sections for 36 situations under different rainfall probabilities and antecedent soil moisture conditions was simulated (Table 5).

From an analysis of the data, the conclusion was drawn that the impact of landscape urbanization on water concentration is also determined by rainfall intensity and antecedent soil condition. Figure 4 shows the flood peak discharges at three different rainfall probabilities at AMCI, AMCII, and AMCIII. The flood peaks in 2000 took place generally 1–2 h earlier than in 1980. This indicates that urbanization could lead to a quicker run-off confluence and higher flood discharge. These will bring more difficulties to flood emergency management.

The Case of Shanghai

Shanghai is the centre of the Yangtze River delta metropolis, and its urbanization level is the highest of the provinces in China. In 2000 it reached 88.31% (the ratio of urban population to total population) and landscape urbanization (not including county or borough data) stood at 86.56%.

Table 5. 24 h flood peak discharge of designed storms with different landscape urbanization scenarios in the Buji River basin (× 10^4 m^3)

		Antecedent soil moisture conditions			
Rainstorm frequency	Year	AMCI	AMCII	AMCIII	AMCI → AMCIII
1%	1980	1058.8	1491.7	1756.3	39.7%
	1988	1167.6	1561.3	1802.0	35.2%
	1994	1254.5	1606.6	1821.8	31.1%
	2000	1270.2	1618.1	1828.4	30.5%
	1980 → 2000	16.6%	7.8%	3.9%	—
2%	1980	887.5	1302.9	1562.9	43.2%
	1988	986.7	1370.1	1607.7	38.6%
	1994	1069.5	1415.3	1626.5	34.2%
	2000	1085.5	1425.1	1633.0	33.5%
	1980 → 2000	18.2%	8.6%	4.3%	—
5%	1980	648.0	1041.3	1296.4	50.0%
	1988	739.0	1103.1	1335.3	44.7%
	1994	816.1	1147.0	1354.6	39.8%
	2000	830.3	1155.6	1358.9	38.9%
	1980 → 2000	22.0%	9.9%	4.6%	—
5% → 1%	1980	38.8%	30.2%	26.2%	—
	1988	36.7%	29.3%	25.9%	—
	1994	34.9%	28.6%	25.6%	—
	2000	34.6%	28.6%	25.7%	—

Note: In the table, 1%, 2% and 5% are rainfall frequency; 1980 → 2000 represents the comparative change of peak discharge when the condition of land use in 1980 changed into that in 2000; AMCI: low soil moisture; AMCII: moderate soil moisture; AMCIII: high soil moisture; 5%?1% refers to the comparative change when the rainstorm frequency changes from 5% to 1%.

Data and Methods

Surface data, hydrological data, and meteorological data were used to study the process of run-off yield and concentration triggered by landscape urbanization in Shanghai. The district of Shanghai has been adjusted several times after 1949, which adds difficulty to the study. Ten urban zones (Yangpu, Hongkou, Zhabei, Putuo, Changning, Nanshi, Jingan, Luwan, Xuhui, and Huangpu: 280 km^2 in total) were selected as the study area. Based on local land use data for 7 years, the change of land use structure from 1950 to 2001 was simulated. The meteorological data used in this paper consisted of precipitation from May to September in 1991 and maximum 24 h storm precipitation in 1962. As the hydrological data were inadequate to directly explore the change of run-off yield in urban areas of Shanghai using the run-off yield formula, the authors had to simulate the situation in other ways. First, based on the water yield from different land cover types of the Taihu Lake basin with each rainfall from May to September 1991 (Gao & Wen, 2002), the water yield per unit area for different land use types was calculated using the following formula:

$$R = R_{\text{total}}/(A^{*}P) \tag{3}$$

where R stands for water yield from different land use types per unit area, R_{total} refers to the gross water yield in the Taihu Lake basin, P refers to the precipitation from May to September 1991, and A refers to the area of each land use type. Considering Shanghai as a part of the Taihu Lake basin, the water yield of various land use types in Shanghai was estimated with the data gained using Equation (2). In addition, the run-off depth of

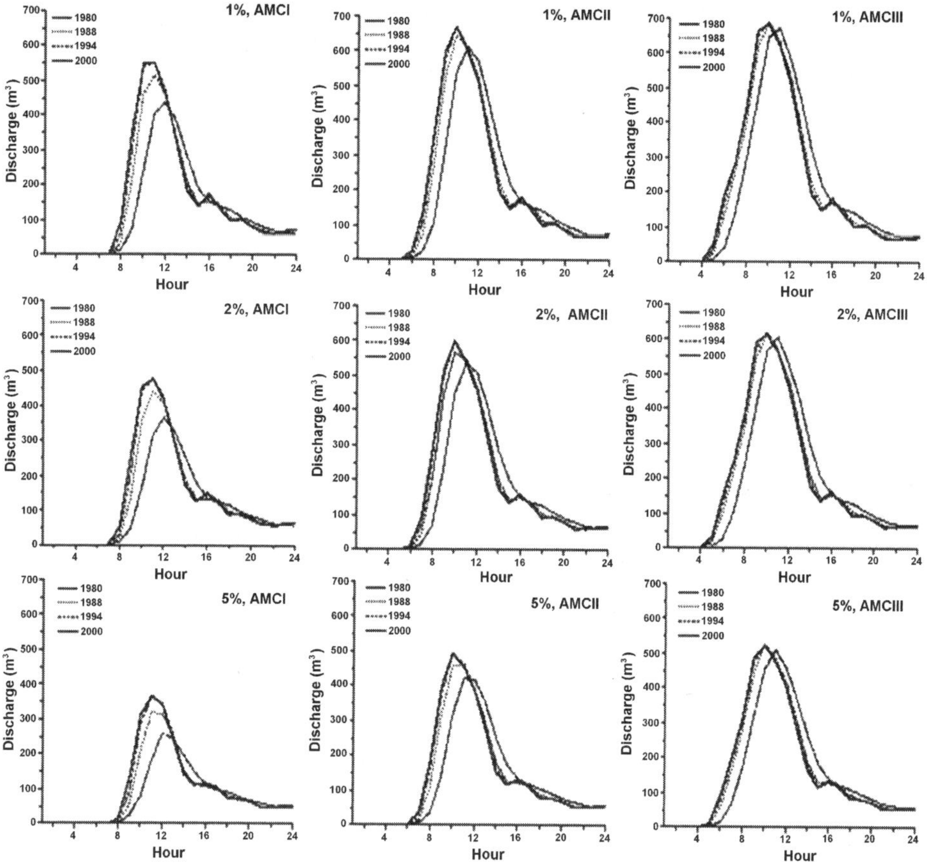

Figure 4. 24 h flood peak discharge under the designed storms (1%, 2%, 5%) and different landscape urbanization scenarios (1980, 1988, 1994, 2000).

different land uses in Shanghai was estimated (Table 6). Although there is an error of absolute measurement with this estimation method, the emphasis of this paper is on the trends in water yield in time series caused by land use changes, and therefore the impact of this error can be ignored.

Results and Analyses

The land use structure change simulated above (1950–2001) shows that the area of construction land (namely, urban land) rose to 79.97% in 2001 from 59.93% in 1950, and the area of water decreased to 0.41% in 2001 from 1.71% in 1950. As a result, the water storage capacity decreased substantially.

The urban area of Shanghai began to experience a surface subsidence from 1921 and a 'disk-like depression' at the bottom of the urban centre was formed (Yan *et al.*, 2002). Without taking drainage into account, the water level measured in August 1962 in Huangpu Park, Shanghai, was taken as the benchmark water level for this research. The benchmark water level (4.76 m) should be considered as a synthesized water level resulting from the

Table 6. Estimated run-off depth (in mm) for different types of land use in the Shanghai urban area

	Land Use			
Year	Construction land	Farmland	Water	Paddy-field
1986	711.4	657.2	468.2	387.6
1996	784.8	683.2	468.2	347.5
Average	748.1	670.2	468.2	367.6

natural water level and the social water level. The influence of sea-level change and some other factors was ignored here. To assess the influence of land use on water level using a quantifying method, it was assumed that the surface subsidence from 1950 to 2001 was the same as in 1962, because there are relatively complete data for this year. First, with the maximum 24 h storm precipitation in 1962, the difference between water yield in 1962 and in any other year was calculated; second, the annual area of urban water bodies was divided by the difference in water yield; and third, the difference in water level was added to the original level of 1962, so that the approximate water-level change from 1950 to 2001 was estimated (Figure 5) (here, the shore gradient change is ignored). This result shows that landscape urbanization results in the rise of the social water level. When the natural water level is overlapped by the social water level and the problem of the so-called 'little flood, serious consequence' appears, it turns into an accelerative process of heightening the flood risk in metropolitan areas.

Integrated Management Strategies to Alleviate Flood Risk

Based on the above study, the conclusion can be drawn that water resource protection and flood prevention have the same importance and they are mutually dependent. In regions of water shortage, especially coastal regions, overusing groundwater has led to sea water

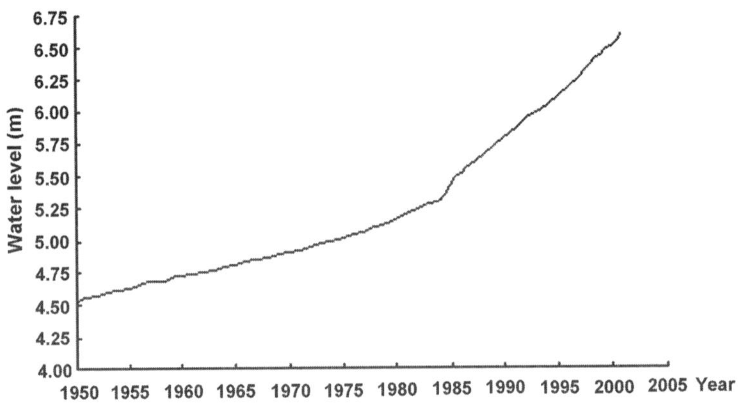

Figure 5. The estimate of water level affected by land use in the 'disk-like depression' of the Shanghai urban area.

intrusion and elevation loss, which further adds to the relative flood level and aggravation of flood disaster. In regions with abundant water resources, wetland is decreasing on account of land use/cover change caused by accelerating landscape urbanization. This weakens the regional flood control ability, raises the social water level, and increases flood risk. Therefore, in metropolitan areas, flood risk control should be strengthened and disaster mitigation should be combined with water conservancy development, so that a regional integrated risk management pattern can be formed.

Adjusting Land Use Structure and Pattern, Balancing the Natural Water Level and the Social Water Level

Floodwater in a metropolis is a product of the natural flood level and the social flood level. One of the important strategies is to balance these levels (Shi *et al.*, 2002b, 2003). However, implementing this strategy is still a difficult problem. According to the two cases above, which illustrate the role of landscape urbanization in increasing surface water yield and concentration comparatively, it is obvious that we should adjust improper land use structures and patterns. Then, the balance between the natural flood level and the social flood level can be retained, the surface seepage rate can be increased, and the run-off coefficient can be reduced. With reference to the authors' successful experience in Dongting Lake and testing in the Buji River basin of Shenzhen and Shanghai, it is found necessary to keep the percentage of various types of 'ecological land' above 40% in the urbanizing basin,[2] in which vegetation should not be lower than 50% in the regions with a slope over 10°, and urban vegetation should not be lower than 30% in the regions with a slope under 10°. These measures can decrease the run-off coefficient by 30–40%. In metropolitan regions, especially on downstream deltas, such as the Yangtze River delta and the Pearl River delta, in which the relative gradient is lower than 1/1000, urban wetlands and water areas should not be lower than 10% and the water area should reach about 2%. For those metropolises with higher population density, it is a wise policy to develop satellite towns.

Constructing a Flood Risk Transferring and Sharing Mode

Flood disaster management in metropolises is an integrated effort involving insurance, water conservancy, and the development of agricultural, economic, and civil affairs. Therefore it calls for co-operation among them in transforming an original mode of government and society into a new one of government, society, and insurance companies (Zhou & Shi, 1999).

Government. Government is responsible for decision making and macro-level risk management, and hence its participation is very important in flood risk management. However, the decision making of government in flood risk management is limited by economic and political conditions. It is urgent for government to clearly delineate the responsibility and obligation of each management sector in order to avoid overlap, conflict, and discontinuity. Government should also develop flood prevention policies and water resources insurance. Meanwhile, it must establish policies and regulations for flood risk management to confirm the rights and obligations of victims and beneficiaries, and set up risk-sharing mechanisms among different groups.

Society. The function of society is mainly performed by non-governmental organizations and communities, whose work generally involves propaganda, education, assisting government to carry out policies, and organizing social donations. Providing social relief is extremely important in coping with flood disaster and is indispensable for flood risk management. In China, there is obvious advancement in social relief, although problems still exist.

Insurance companies. According to certain flood prevention standards developed by government, insurance companies should actively expand flood risk insurance and enrol more community members. Technical support for flood prevention should also be strengthened to reduce losses caused by floods to the minimum level and further reduce the loss ratio of insurance companies.

Conclusions

Although the authors have determined the optimal mix of ecological land, wetlands, and water bodies for urbanizing areas, especially in the delta plains, it is necessary to consider how to plan the spatial pattern of ecological land. The study points to two critical considerations in this context.

(1) Adjust the land use structure and pattern in order to control and prevent urban floods. Conserving ecological land over 40% of the urbanizing region and maintaining wetlands and water bodies over 10% could limit the increase in flood disaster in China.
(2) Establish an integrated risk management plan to cope with flood disaster in the future. Flood risk insurance as an effective measure will enhance flood prevention and mitigation, and improve the efficiency and benefit of disaster mitigation. Integrated risk management should shift from the government–society paradigm to the government–society–insurance paradigm, which will help facilitate regional sustainable development.

Acknowledgements

This project was supported by the National Science Fund for Distinguished Young Scholars (no. 40425008), the International Institute for Applied Systems Analysis, and the Disaster Prevention Research Institute. The authors thank Professor Chennat Gopalakrishnan of the University of Hawaii at Manoa and Professor Norio Okada of Kyoto University for reviewing this paper.

Notes

1. Social water level: an additional water level caused by irrational and unsafe human activities such as human occupation of wetlands and water areas, deforestation, etc.
2. Ecological land: the land has clear ecological value and includes farmland, garden plots, forest, bush, grass, wetlands, and water.

References

Brun, S. E. & Band, L. E. (2000) Simulating runoff behavior in an urbanizing watershed, *Computers, Environment and Urban Systems*, 24, pp. 5–22.

Fu, C. (2001) How much can the Asian monsoon be modified by human-induced land cover changes? Land use and East Asia monsoon, *Abstract of Global Change Open Science Conference* (Amsterdam: Congress Holland).

Gao, J. & Wen, Y. (2002) Impact of land use change on runoff of Taihu basin, *ACTA Geographical SINICA*, 57, pp. 194–200, (in Chinese).

Guangzhou Institute of Geography Science (1983) *Landform of Shenzhen City* (Guangzhou: Guangdong Science and Technology Press).

Hall, M. J. (1984) *Urban Hydrology* (London: Elsevier).

Lambin, E. F., Baulies, X., Rockstael, N., Fischer, G., Krug, T., Leemans, R., Moran, E. F., Rirdfuss, R. R., Stao, Y., Skole, D., Turner, II, B. L. & Vogel, C. (1999) Land use and land cover change implementation strategy. International Geosphere–Biosphere Programme Report 48 and International Human Dimensions Programme on Global Environmental Change Report 10, Stockholm.

Lewis, M. J., Singer, M. J. & Tate, K. W. (2000) Applicability of SCS curve number methods for a California oak woodlands watershed, *Journal of Soil and Water Conservation*, 20, pp. 226–230.

Liu, C. & He, X. (1996) *Water Problem Strategy for China's 21st Century* (Beijing: Science Press).

Liu, S. (2004) Science and technology innovation to resolve the water resource issue, *Chinese Education Newspaper*, 31 May, p. 4.

Lu, J. (1990) The efficient use of remote sensing in hydrology model, *Hydrology*, 6, pp. 9–14.

Office of State Flood Control and Drought Relief Headquarters (1997) *Flood and Drought in China* (Beijing: China Water Power Press).

SCS (1972) *National Engineering Handbook* (Washington, DC: US Government Printing Office).

Shenzhen Water Resources Bureau (1994) *The Plan of Urban Flood (Tide) Control in Shenzhen* (Shenzhen: Ministry of Water Resources).

Shi, P. (1997) The water crisis in China: problems and strategies, *Stratagem and Management*, 6, pp. 40–47.

Shi, P., Gong, P., Li, X., Chen, J., Qi, Y. & Pan, Y. (2000) *The Methods and Practice of Land Use and Land Cover Change Research* (Beijing: Science Press).

Shi, P., Wang, J., Yang, M., Wang, Y., Ding, Y., Zhuo, L. & Zhou, J. (2002a) Integrated risk management of flood disaster in metropolitan regions of China (I)—to balance flood disaster magnitude and vulnerability in metropolitan regions. Paper presented at second annual International Institute for Applied Systems Analysis–Disaster Prevention Research Institute (IIASA–DPRI) meeting, 29–31 July. Vienna, Austria.

Shi, P., Song, C. & Jing, G. (2002b) Strengthening the study of land use/cover change and its impact on eco-environmental security—the trend of the study of the dynamics of the human–nature system based on the 'Global Change Open Science Conference 2001' in Amsterdam, *Advances in Earth Sciences*, 17, pp. 161–168 (in Chinese).

Shi, P., Yuan, Y., Ge, Y., Zhou, J. & Yang, M. (2003) Integrated risk management of flood disaster in metropolitan regions of China (II)—to adjust land use structure and pattern in metropolitan regions. Paper presented at third annual International Institute for Applied Systems Analysis–Disaster Prevention Research Institute (IIASA–DPRI) meeting, 3–5 July, Kyoto, Japan.

United Nations Educational, Scientific, and Cultural Organization (1974) Hydrological effects of urbanization, *Studies and Reports in Hydrology*, 18, Paris.

Wang, Y. (1989) Hydrology model of remote sensing and its application, *Hydrology*, 5, pp. 20–24.

Yan, X., Gong, S. & Zeng, Z. (2002) Relationship between building density and land subsidence in Shanghai urban zone, *Hydrogeology and Engineering Geology*, 6, pp. 21–25.

Zhang, L. & Shi, P. (1994) Establish system dynamics of human and earth and strengthen integrated research in environmental and ecological problems, *Fund of Science in China*, 3, pp. 158–160.

Zhou, W. & Shi, P. (1999) Advances in flood risk management study and the primary research on flood risk management models in China, *Journal of Natural Disasters*, 8, pp. 62–72.

Rapid Economic Assessment of Flood-control Failure along the Rio Grande: A Case Study

ZHUPING SHENG, ALLEN STURDIVANT, ARI MICHELSEN & RON LACEWELL

Introduction

Reducing public risks associated with severe weather, natural disasters, and other *force majeure* can be a key factor in a region's ability to sustain economic development and growth (International Federation of Red Cross & Red Crescent Societies, 2002). Threats from such natural hazards can impose extraordinary economic and financial costs to society when they occur. Though imperfect, disaster mitigation and preparedness are known to pay—in human, economic, and environmental terms. One of the most cost-effective ways to reduce the risk of disaster is prevention. In this study, we developed an innovative geographic information system (GIS) integration method, i.e. joint application of land use, appraisal district property values, localized and modified flood inundation data, and GIS coverage, for delineation of flood inundation areas, identification of land use types, and extraction of property values, to provide the US Section of the International Boundary and Water Commission (USIBWC) the economic information needed to support its federal funding request for rehabilitation of levees and other flood-control infrastructure along the Rio Grande. Current flood-control infrastructure has deteriorated and may be subject to failure, thereby putting these regions at risk.

Over 5 million people live in the cities and towns in the USA and Mexico along the Rio Grande where extensive housing, commerce, industry, tourism, and irrigated agriculture exist. For example, a total of 684 722 people lived in El Paso County, Texas, in 2002 (City of El Paso, 2004a). El Paso Central Appraisal District data showed that the total property valuation for 2003 was $18.2 billion (City of El Paso, 2004b) (all prices in US dollars). In addition, El Paso County is home to nearly 1700 agribusinesses contributing $7.3 billion to the economy of the area each year (Agribusiness Council of El Paso, 2000). The City of El Paso Building Services Department issued $520 million in building permits in 2003 alone (City of El Paso, 2004b). Although only a portion of the population and property are at risk of flood damage, with the population expected to double within 50 years (Texas Water Development Board, 2002) a portion of the projected property development and agricultural production will be in the flood probe areas and will also be at risk of flood damage. Therefore, without timely and appropriate rehabilitation of levees and other flood-control infrastructure, people along the border could face a greater risk of disaster due to failure of ageing flood-control projects.

The USIBWC constructed and is responsible for maintaining a series of flood-control levees and other water-control infrastructure along the Rio Grande for flood protection on the US side of the USA–Mexico border. In response to observed levee degradation, USIBWC engineers evaluated the structural integrity of the levees and the flood-carrying capacity of the structure along the Rio Grande during 2001–03 (IBWC, 2003, 2004a; Robinson, 2004a). The assessments suggested some levee segments are structurally deficient and IBWC hydraulic analyses indicated flood-conveyance capacities are below original project-design criteria (Moehlig, 2004). The results of those evaluations indicate that structural rehabilitation is necessary to restore the flood control to its original design capacities (Stefanov, 2004).

Recent flood events along the Rio Grande, combined with USIBWC's structural integrity evaluations and the Office of Management and Budget's requirement that agency-requested costs for flood-control infrastructure be economically justified, highlighted the need for the agency to promptly assess the potential economic impacts of a levee breach and/or failure of flood-control infrastructure. Because of USIBWC's short time frame (estimates were needed within 2 months to accompany the agency's budget request), the large geographic extent of the river levee and floodway levee system, and the absence of any previous economic evaluation (IBWC, 2004b), an innovative rapid assessment approach was required. This approach relies heavily on integrating a variety of GIS/geospatial information to identify flood inundation areas and land use, update existing maps and data, and develop estimates of potential property value damage from the failure of four USIBWC flood-control projects on the Rio Grande. Multiple economic assessment methods were employed to estimate both urban property and agricultural production losses caused by flood inundation (e.g. US Water Resources Council,

1983; Lacewell *et al.*, 1989, 1995; Lacewell & Freeman, 1990; US Army Corps of Engineers (USACE), 2003, 2004; Freeman, 2004; Robinson, 2004b; US Bureau of Reclamation, 2004). A nationally standardized risk assessment software program, i.e. HAZUS-MH (multi-hazard), which contains models for estimating potential losses from earthquakes, floods, and hurricane winds (Federal Emergency Management Agency (FEMA), 2004a), was not used because the necessary detailed hydrology and digitized cross-section data for the study areas were not yet available. The methods developed for use in this study could be applied in other areas where standardized models and data do not yet exist.

Project Areas

This study examines the economic losses avoided from flood protection and is limited to the expected flood-protection benefits for the US side of four USIBWC flood-control projects along the Rio Grande on the US side. These projects extend in segments from Caballo Dam in south-eastern New Mexico, and continue downstream 2145 km or 1333 miles along the Rio Grande and end near the Gulf of Mexico in southern Texas. The four projects are the Rio Grande Canalization (RGC) project (209 km/130 miles), the Rio Grande Rectification (RGR) project (150 km/93 miles), the Presidio Valley Flood Control (PVFC) project (24 km/15 miles), and the Lower Rio Grande Valley Flood Control (LRGVFC) project (164 km/102 miles along the Rio Grande and 441 km/274 miles including interior channel levees). These are shown in Figure 1 with the flood-control capacity and floodway characteristics of the four project areas summarized in Table 1.

RGC Project

The RGC project's purpose is to provide flood protection against a 100-year flood and convey water to American Dam, where the USA diverts its allocated amount and passes Mexico's allocation through the Rio Grande in accordance with the 1906 convention. The RGC project infrastructure, which is entirely in the USA, includes about 209 km (130 miles) of levees from Percha Diversion Dam (3.2 km/2 miles below Caballo Dam) in New Mexico to American Dam in El Paso, Texas, on the USA–Mexico border. Of these levees, it is estimated that 16 km (10 miles) would be overtopped by water associated with a 100-year flooding event and an additional 97 km (60 miles) would be subject to encroachment by flood water (IBWC, 2003; Robinson, 2004a). Current flood-design capacities, as defined by the IBWC, are set at 629 cubic metres per second (cms) or 22 200 cubic feet per second (cfs) at Leasburg Dam and/or 396 cms (14 000 cfs) at American Dam (IBWC, 2004a) (Table 1). Within the project area, Hatch and Mesilla in New Mexico, and west El Paso in Texas, are highly populated areas. Therefore, properties near the river in these populated areas would be subject to flood damage if the RGC project's levees failed. On the other hand, agricultural production, which would also be subject to flood damage, is primarily located within Doña Ana County, New Mexico.

RGR Project

The RGR project is designed to protect the El Paso and Ciudad Juarez area from waters associated with a 100-year flood in accordance with the 1933 convention (IBWC, 2004a). The RGR project infrastructure includes 150 km (93 miles) of river levee between American Dam in El Paso and Fort Quitman (i.e. along El Paso and Hudspeth counties on the US side) in the westernmost reach of Texas. Originally, 249 km (155 miles) of levees

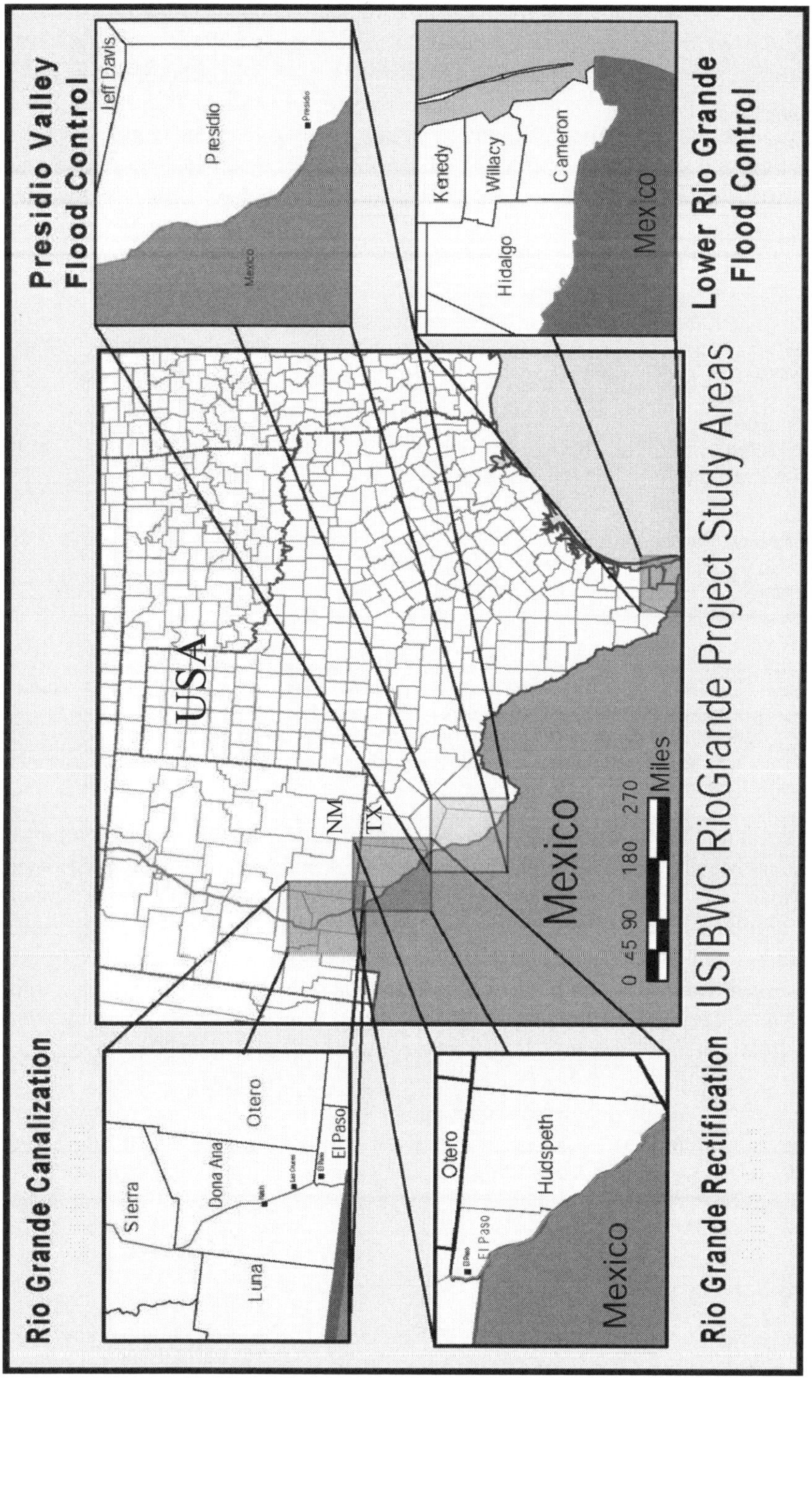

Figure 1. Location of IBWC flood-control project areas.

Table 1. Summary of selected characteristics of USIBWC projects (US side only), 2004

Characteristic	Project: RGC	RGR	PVFC	LRGVFC	Total
Flood information					
Flood design	100-year	100-year	25-year	500-year	—
Freeboard (m)	0.61	0.61	1.22	0.91	—
Designed-flood flow (cubic metres per second)	629/396	312	102/1189	7079	—
At location	At Leasburg Dam/at American Dam	El Paso, TX	Above/below Rio Conchos	Rio Grande City, TX	
River floodway					
Total kilometres of levee	209	150	24.4	164	547.4
Kilometres at risk of being overtopped	16	19	2	61	98
Kilometres subject to encroachment	97	61	2	103[a]	263
Interior floodway[b]					
Total kilometres of levee	0	0	0	277	277
Kilometres at risk of being overtopped	0	0	0	3.2	3.2
Kilometres subject to encroachment	0	0	0	39[a]	39

[a] Note for the LRGVFC project, values based on river levees having 0.91 m (3 foot) of freeboard and floodway levees having a 0.61 m (2 foot) freeboard.
[b] These values are summed for the left and right levees (i.e. with a downstream flow).
Source: IBWC (2004a), Jim Robinson and Albert Moehlig with the IBWC, and data calculations of project personnel.

were present, but the RGR project, which was authorized by a 1933 convention between the USA and Mexico, reduced the amount as the USA and Mexico sought to rectify and stabilize their international boundary (IBWC, 2004a). Another project (i.e. the Chamizal Convention) further reduced the inventory of this reach to the current 150 km (93 miles). Of this current total, it was estimated that 19 km (12 miles) would be overtopped by water flow associated with a 100-year flood event, while an additional 61 km (38 miles) would be subject to encroachment by flood water (IBWC, 2003; Robinson, 2004b). For this project area, the IBWC defines such flooding as a flow rate of 312 cms (11 000 cfs) measured at American Dam in El Paso (Table 1). Central and east El Paso, Texas, are areas within the flood zone that are highly populated. The properties near the river in these populated areas would be subject to flood damage if the RGR project's levees failed. Agricultural production primarily located in the El Paso lower valley and south-west Hudspeth County is almost entirely in the flood protected areas and would also be subject to flood damage.

PVFC Project

The PVFC project's purpose is to protect 13 355 ha (33 000 acres) of agricultural land and the lower part of the city of Presidio, Texas, from waters associated with a 25-year flood in accordance with Title II of Public Law 92-549. The PVFC project infrastructure includes

24.4 km (15.18 miles) of levees stretching upstream and downstream of the tributary entry-point of the Rio Conchos from Mexico into the Rio Grande. Of this total, it was estimated that 2 km (1.25 miles) would be overtopped by water flow associated with a 25-year flood event, while an additional 2 km (1.25 miles) would be subject to encroachment by flood water (IBWC, 2004a; Robinson, 2004a). For this project, IBWC defines such flooding as a flow rate of 102 cms (3600 cfs) above the Rio Conchos, and/or 1189 cms (42 000 cfs) below the Rio Conchos (Table 1). The city of Presidio is the populated area associated with the PVFC project. Only limited properties near the river would be subject to flood damage, however, if the PVFC project's levees failed. Agricultural production which would be subject to flood damage is also very limited when compared to other project areas (Table 2).

LRGVFC Project

The LRGVFC project's purpose is to provide flood protection to urban, suburban, and high value agriculture production in the area, from waters associated with a 500-year flood in accordance with Title II of the National Industrial Recovery Act of 13 June 1933 and the Act of 19 August 1935 (IBWC, 2004a). The LRGVFC project covers 164 km (102 miles) of levees along the main channel of the Rio Grande from Peñitas in Hidalgo County, Texas, to a point 45 km (28 miles) from the Gulf of Mexico. The LRGVFC project infrastructure also includes 277 km (172 miles) of interior floodway levees, which total 441 km (274 miles) of levees. Of this total, it was estimated that 64 km (40 miles) would be overtopped by water flow associated with a 500-year flood event (Moehlig, 2004), while an additional 142 km (88 miles) would be subject to encroachment by flood water (IBWC, 2003). The project's original flood-design capacity of 5295 cms (187 000 cfs) measured at Peñitas, Texas, was increased to 7079 cms (250 000 cfs) (Table 1) measured at Rio Grande City, Texas, following extensive damage caused by Hurricane Beulah in 1967. Penitas, McAllen, Weslaco, and Brownsville in Texas are the populated areas. Properties near the main channel of the Rio Grande and interior floodway levees in these populated areas would be subject to flood damage if the LRGVFC project's levees failed. Agricultural production, which would also be subject to flood damage, is distributed extensively in both Hidalgo and Cameron counties of Texas (Table 2).

Table 2. Land uses (in ha) within the revised flood inundation areas for four USIBWC projects (US side only), 2004

	Project area				Total area	Percentage of total area
Category	Doña Ana County (RGC)	El Paso County (RGC and RGR)	PVFC	LRGVFC		
Agricultural	1005	953	309	30 613	32 880	87.60
Residential	743	1070	129	1 310	3 252	8.67
Commercial	26	1117	0	245	1 388	3.70
Industrial	0	13	0	0	13	0.03
Total area	1775	3153	439	32 167	37 533	100.00

Assessment Methods

A combination of GISs was used to integrate multiple data sources and the different economic assessment methods for urban developed properties and agricultural production. Four levels of analysis were conducted. First, flood inundation areas were identified using FEMA inundation maps and GIS coverage with modification for levee failure conditions. Second, large scale land use types and their acreages were determined by integrating many types of maps and imagery through GISs, including satellite images, multi-resolution land use coverage (MRLC) maps, orthophotography, and appraisal district data. Third, properties at risk were identified and their values were then assessed. Fourth, flood damage and economic losses avoided by flood control projects were estimated for both developed properties and agricultural crops using published factors (Figure 2).

For urban areas, two assessment methods were used. Where individual property values were available, the authors overlaid the property parcels coverage on the identified flood

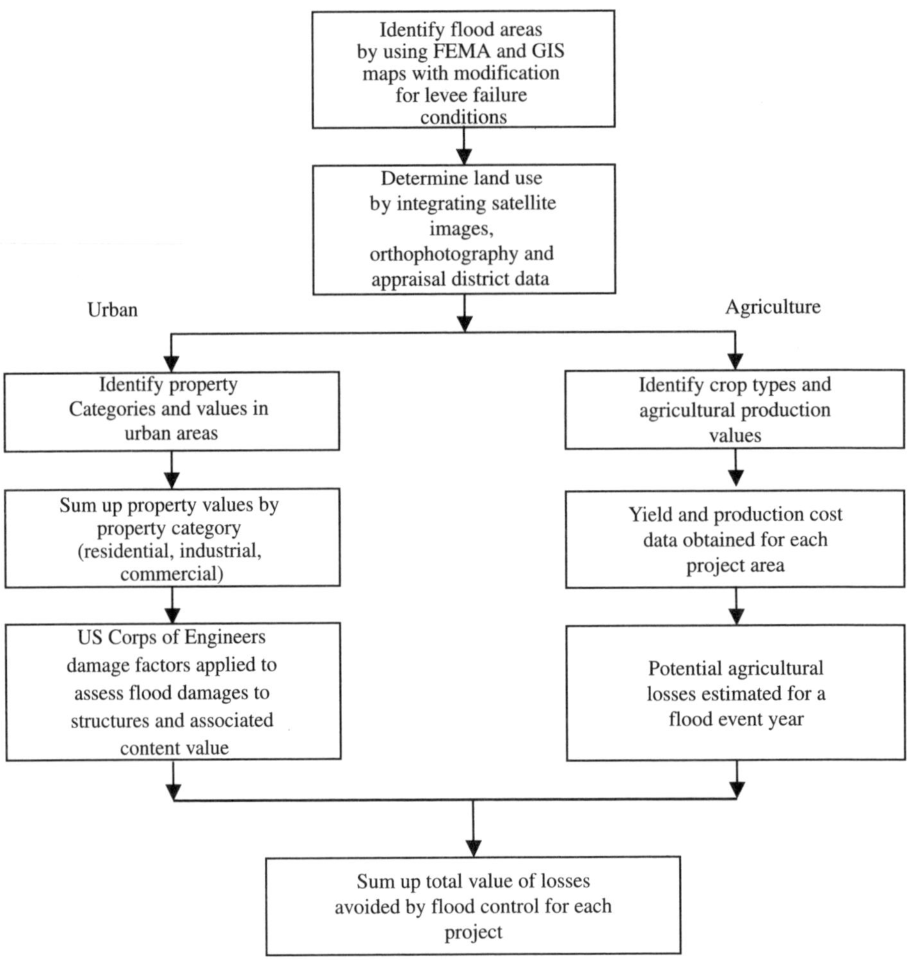

Figure 2. Flow chart of the integrated assessment method.

inundation areas to identify all the individual properties and their values within flood zones (Figure 3). These property values were summed for each category, i.e. residential, commercial, and industrial. Where individual property values were unavailable, such as for the LRGVFC project, representative land uses for the area were estimated and extrapolated to other similar areas. Once the flood inundation property values by the type of development were established, USACE (2004) damage factors were applied to estimate flood stage damage to structures and associated content value damage. Integration of a variety of maps and associated geo-referenced data sets provided a powerful tool for a rapid economic assessment of flood-control losses avoided for those four USIBWC project areas. Because of differences in data availability, several iterations and some unique procedures were required for each of the project areas.

For agricultural production estimates a composite crop use pattern for each project area was determined using several sources of agricultural statistics (Lacewell *et al.*, 1995; Texas Department of Agriculture, 1998; Texas Cooperative Extension, 2003; McGuckin, 2004). Yield and production cost data were also obtained from several sources to determine agricultural production values (Lacewell *et al.*, 1995; Texas Cooperative Extension, 2003; McGuckin, 2004) for each project area. For the LRGVFC project area, an economic model, the agricultural benefit estimator (ABE) developed specifically for this region by Lacewell & Freeman (1990), was used to assess flood-control losses avoided. Potential agricultural losses were estimated by assessing the difference in net revenues that producers would realize between (1) a natural, non-flooded crop production and harvest year and (2) revenues in flood event conditions.

Integration of Data through GISs

To assess impacts of floods on surrounding areas along the four different project areas, a GIS integration method was developed for delineation of flood inundation areas, identification of land use types, and extraction of property values. Many sources of digital and hard copy data, including FEMA flood inundation maps, FEMA flood insurance rate maps (FIRMs) for Presidio County (FEMA 1985, 2004b), digital orthophoto quarter quadrangles (DOQQs) at a 0.46 m (1.5 feet) to 1 m (3.3 feet) resolution from 1993 to 2002 depending on update availability, MRLC maps at a 30 m resolution for the year 1992 provided by Texas A&M University Spatial Sciences Lab (SSL), and remote sensing images (LandSat 7 ETM +) at a 15–30 m resolution for the year 1997 archived by the University of Texas at El Paso, were used to identify types of land uses or a classification set provided by IBWC, and potential flood inundation areas (Figure 4).

Once land uses were identified within the potential flood inundation areas, the acreage of agricultural production (row crops, pasture, orchard) and residential, commercial, and industrial purposes was also estimated based on the higher resolution land use maps. Multiple sources of data were used to verify the accuracy of information across sources. Wherever data were available, the most recent data/images were used in this assessment. Much development has occurred however, since the FEMA and US Geological Survey (USGS) maps were created. In some cases, this was addressed by using more recent property land use and value data from appraisal offices. For PVFC and LRGVFC areas, the land use figures are from 1992, and do not include significant development.

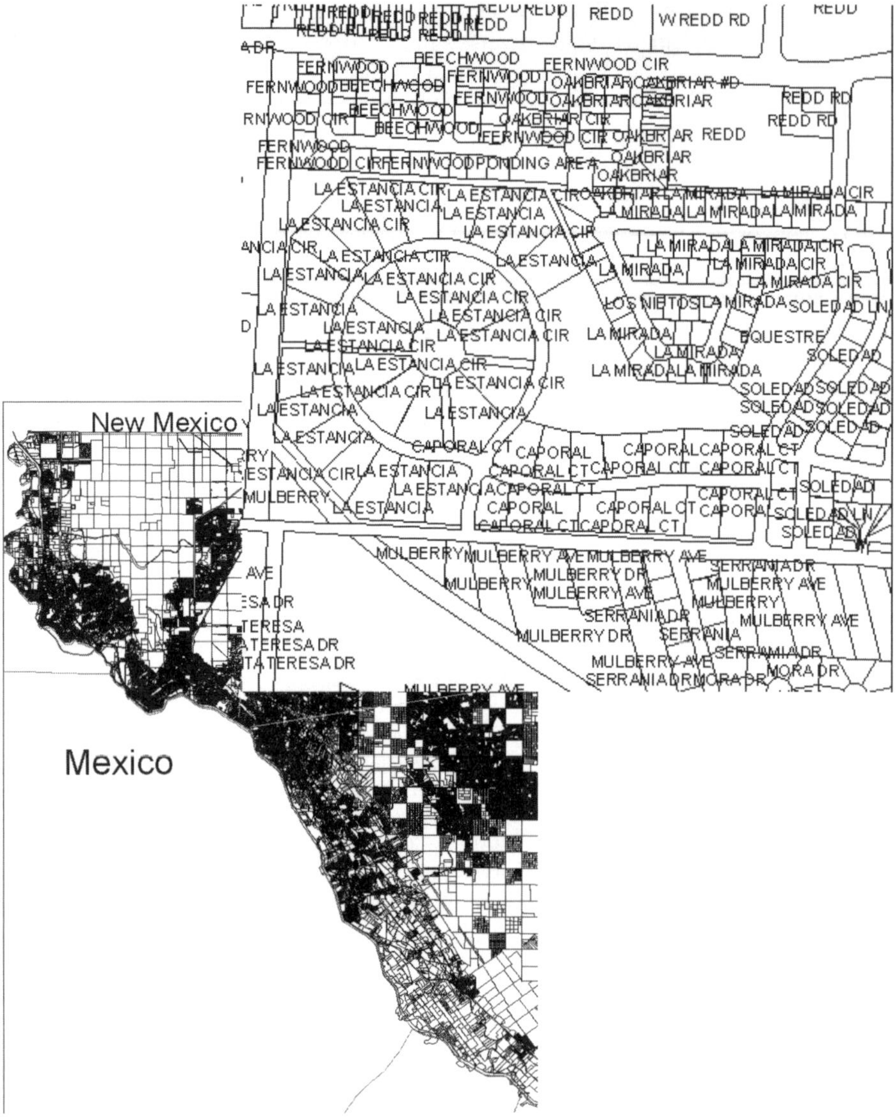

Figure 3. Urban property parcels used for economic assessment.

Identification of, and Adjustments to, Potential Flood Inundation Areas

A variety of GIS and hard copy data were used to identify land uses and potential flood inundation areas. FEMA FIRMs were used to identify flood inundation areas, which are based on rainfall events (FEMA, 1985, 2004b). As such, the total flooded area estimated by FEMA includes several disconnected areas such as low lying land spaces which are disconnected from the river by a rise in the land-surface elevation or higher elevation arroyos (intermittent stream flow) that would not be flooded with a structural failure of flood-protection infrastructure along the Rio Grande. Thus, to more accurately estimate

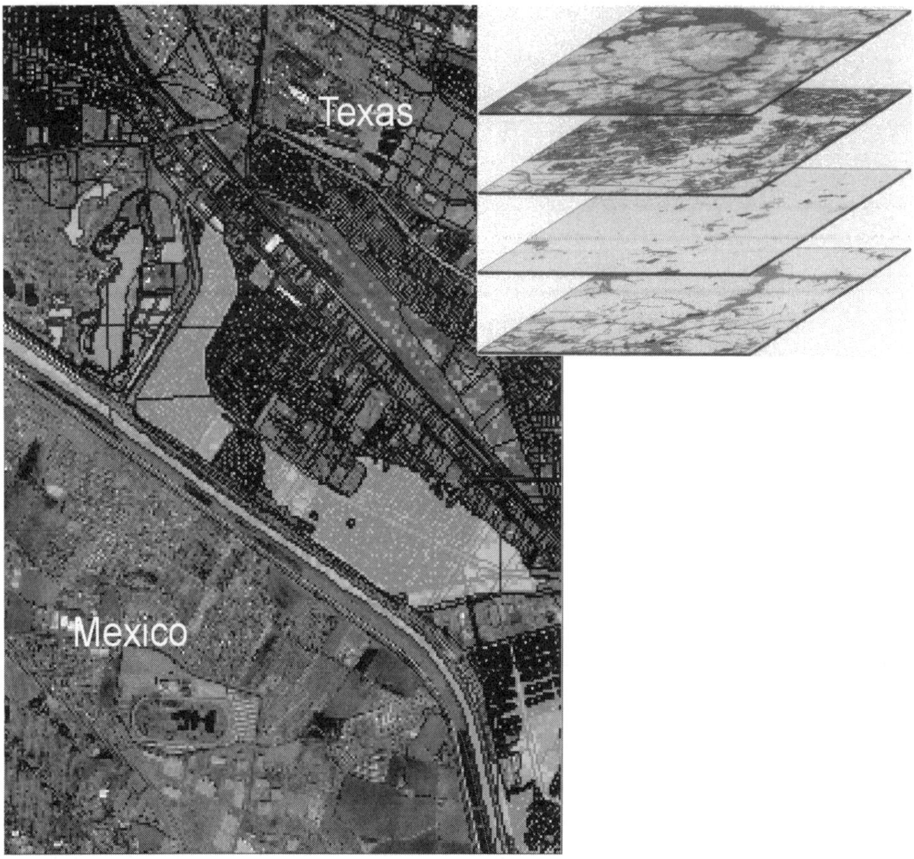

Figure 4. Assessment of flood damage using GIS integration of FEMA FIRMs, DOQQ and MRLC images, and a property parcel map.

the appropriate floodplain area associated with an infrastructure failure along the river, the disconnected and high elevation precipitation-associated flood areas in the FEMA map were excluded, thereby resulting in a revised FEMA floodplain map used in this analysis. For Presidio County, Texas, FEMA maps were available only for the unincorporated area. Working with the flood area from the FEMA map, insurance industry maps were used to define the section of the city likely to flood.

MRLC data were then used to identify land uses within the flood inundation areas. The same land use categories and types were applied to all four project areas. Urban land use was separated into residential, commercial, and industrial types. Agricultural land use represents all enterprises (e.g. row crops, vegetables, citrus, pasture, etc.) grown in each project area. Land use classifications were then cross-verified with other sources such as DOQQs and satellite images to assure correct delineation of flood inundation areas and correct classification of land use. For urban property, a major source of verification on the individual property level was property assessment records. There are instances when certain individual properties lie partially in the flood inundation area. In these cases, the standard procedure was to include the 'partially in' properties while defining the revised floodplain area.

Determining Urban Property Values and Damage

The standard protocol applied in this analysis is referred to as the property damages avoided (PDA) method, which has historically served as the USACE initial screening criterion when considering flood-control benefits (Shabman *et al.*, 1998). The PDA method takes a practical approach as it measures flood damage based on estimated property restoration and repair costs. These are assumed to be consistent with individuals' willingness to pay for flood protection. In a study for the USACE, Shabman *et al.* (1998) compared the PDA approach with both the contingent valuation and hedonic property methods for assessing flood-control benefits. When comparing consistency, reproducibility, and credibility across the three methods, the authors did not find significant justification to deviate from the PDA benefit-estimation method, as long as its limitations were recognized.

Although the PDA method was applied for all project areas, the specific approach and tasks involved with obtaining, analysing, and managing the requisite data varied slightly across the four project areas. As expected, there were variations in data sources, format, type, quality, quantity, etc.

The value and damage factors (i.e. for structure, contents, inventory, and equipment) applied in determining damage are adopted from work analysing stage-damage curves used by the USACE (Freeman, 2004; USACE, 2004). The appraisal values for urban land use structures are key determinants in the potential damage estimates and were separately determined for each project area. The assumed values were calculated from data obtained from various local data sources (e.g. appraisal districts).

Content values apply only to residential structures and associated improvements, while inventory and equipment values apply to commercial and industrial structures. The residential contents value is assumed to be 50% of the appraised structural value. This proportion was obtained from the USACE–Institute for Water Resources (USACE–IWR) (Shabman *et al.*, 1998). This value, for a 2-foot (0.61 m) flood inundation, is considered conservative and is commonly used by others (e.g. FEMA, USACE) with respect to single-family residences, without basements (Freeman, 2004).

Similarly, inventory and equipment proportions remain consistent across both commercial and industrial structures and are assumed to be 73% and 63%, respectively, of the appraised structural value. Since there are no IWR generic curves available for the urban land uses, an alternative method was required. Specifically, historical stage-damage relationships obtained from the USACE Galveston District for a 2-foot inundation were analysed by Standard Industrial Classification divisions (USACE, 2003). The damage factors are based on detailed survey data obtained following flood events. Urban land use damage factors (or proportions of dollar values) are applied to structure, contents, inventory, and equipment values for residential, commercial, and industrial structures.

Residential. For residential structures, the process presented employs the residential stage-damage curves to compute damage–benefits. Generic stage-damage curves from the USACE are used to establish the expected damage associated with 2 feet of inundation. The damage factor for single-family residential structures, without basements, is used for this analysis. For residences, the structural damage factor is 32.1% of the appraised structural value, while the contents damage factor is 17.9% of the contents value.

Commercial and industrial. Commercial and industrial damage mirror each other and are calculated as for residential properties. In addition to structural damage, commercial and industrial properties are assumed to incur two other categories of damage: inventory and equipment. For commercial structures, the structural damage factor is 11.1% of the appraised structural value, the inventory damage factor is 36.0% of the inventory value, and the equipment damage factor is 22.0% of the equipment value (with inventory and equipment value estimated as discussed earlier).

Determining Agricultural Values and Damage

Agricultural production values and potential damage were estimated for each project area taking into account specific crops produced and production values. In Doña Ana County, New Mexico, the county associated with the RGC project, this assessment was done using the FEMA map in conjunction with a property parcel shape file (in GIS format) provided by the Doña Ana Assessors' Office and satellite land use image maps provided by the SSL in the Texas Agricultural Experiment Station (TAES) at Texas A&M University. A composite crop use pattern was determined from 2001 New Mexico agricultural statistics. Yield and production cost data were obtained from the projection for 2004 of New Mexico Cooperative Extension's costs and returns for Doña Ana County, New Mexico (McGuckin, 2004). It was assumed that flood damage to field crops will occur at the most critical period of production; thus, for most crops, a total loss is assumed. This was based on the timing of historical flood events. Alfalfa production was assumed to lose one cutting (i.e. equivalent to 20% of normal yield), while pecan orchards, which are routinely flood-irrigated, were assumed to incur no damage. Crop prices were obtained from the Economic Research Service, US Department of Agriculture (2003), and are referred to as normalized prices. These are the prices used for agricultural products and are designed to eliminate the impact of farm programmes.

El Paso County, Texas, is associated with the RGC (upper valley above American Dam) and RGR (lower valley below American Dam) projects. A similar assessment method as the one used for Doña Ana County was used for estimating potential damage to agricultural production. The land use GIS map provided by the TAES SSL was used to obtain the acreage of row crops and orchards. The crop mix, based on county statistics, includes alfalfa, pima cotton, red chile, wheat, and pecans. Orchards are almost exclusively pecans. Row crop acreage was distributed at proportional levels of the county's crop mix. Texas Cooperative Extension (2003) crop and livestock enterprise budgets were applied to establish expected yield and harvest costs per unit of output. Where there is a yield loss, the costs to harvest are reduced in correspondence with the yield loss.

Presidio County, Texas, is associated with the PVFC project. The land use map of the flood inundation area provides acreage of row crops. There are no orchards in this region. For acreage estimates of agricultural land, discussions with the Presidio County agricultural agent provided insight on the crop mix in the county. Due to long-term drought and limited irrigation water, the only crops grown are alfalfa and other hay. The allocation of total agricultural acreage between the two crops is 80% alfalfa and 20% other hay. Again, Texas Cooperative Extension (2003) crop and livestock enterprise budgets were applied to establish expected yield and harvest costs per unit of output. Where there is a yield loss, the costs to harvest are reduced in correspondence with the yield loss.

Hidalgo and Cameron counties in Texas are associated with the LRGVFC project. This region was handled differently as there is a long history of research related to flood damage for agricultural crops in the region. That is, previous drainage projects of the USACE (1982) involved an agricultural component. In addition, an economic model was previously developed specifically for this region—the ABE model—by Lacewell & Freeman (1990). Potential agricultural losses are estimated by assessing the difference in net gross revenues that producers would realize between (1) a natural, non-flooded crop production and harvest year and (2) a year in which a 100-year flood event occurred, affecting production and/or harvest of crops in the designated floodplain region. A composite acreage approach is utilized for row crops, encompassing the LRGVFC regional crop mix, with some adjustments incorporated to account for those crops not usually planted in the designated floodplain region, as well as for those crops for which the proportion of plantings is greater inside the designated floodplain region than outside that area. Citrus and pasture acreage are considered separately from the composite crop acreage, but in a consistent manner otherwise. The composite acreage is simply the proportion of each row crop that comprises a typical acreage in the floodplain. This typical acreage was developed through focus groups used for the USACE studies on benefits of drainage in this area (Robinson *et al.*, 1992; Lacewell *et al.*, 1995, 1997, 2001). Differences in normal versus flood-affected yields are based on percentage loss in yield developed by focus groups. The data for this region include the expected non-flood yield by crop and by soil type.

Texas Cooperative Extension (2003) crop and livestock enterprise budgets were used in determining per unit harvesting costs. Normalized crop prices were used along with the weighted yield, across relevant soil types, to identify the gross value of the harvested crop (with no quality damage assumed when flooded). The process is the same as for other regions where a damage factor from the ABE model is applied to the gross revenue for losses, but the harvest costs for the yield loss are excluded as a cost. A particular difference for this region compared to the others is the inclusion of additional costs associated with flooding of agricultural land beyond yield loss. This includes field reconstruction, laser levelling, infrastructure repairs, etc. These per hectare values were included in the ABE model (Lacewell & Freeman, 1990) and developed from focus groups (Lacewell *et al.*, 1995).

In summary, reductions in harvest and post-harvest processing costs are accounted for based on the Texas Cooperative Extension (2003) crop and livestock enterprise budgets, effectively reducing the apparent damage gauged by a comparison of gross revenues between the normal and flooded scenarios. The ABE model (Lacewell & Freeman, 1990) was used in the calculation process, incorporating the probabilities of flooding at different times of the year, and adjusting potential loss estimates according to the respective production stages of the various crops. Multiplication of the respective crops' proportion in the estimate by total cropped floodplain acreage results in acreage of each crop. Applying the per hectare losses identified in the above method multiplied by the total respective crop acreage in the estimated floodplain (i.e. the FEMA map less 'disconnected areas') yields the estimated losses for each crop due to flooding.

Results of Economic Assessment

The economic analysis is based on a 2-foot (0.61 m) flood inundation depth and is applicable to only the US side of the border. Further, the reported values only include property damage, and do not consider potential damage for items such as loss of life, infrastructure, etc. Therefore, these results are considered conservative since some more recent urban development is not included, damage from moving water is ignored, and some areas are inundated at much greater depths. Limitations and recommendations are addressed in more detail later in the paper. Aggregate economic losses avoided by all four flood-control projects conclude the article.

Characterization of Land Uses

Using GIS integration, types of land use and acreages for each category within potential flood inundation areas were identified and quantified for the four project areas (Table 2). The total flood inundation area includes 37 533 ha (92 746 acres) across the four project areas, of which 4653 hectares (11 497 acres), 12%, is urban development and 32 880 ha (81 249 acres), 88%, is agricultural. For illustrative purposes, El Paso County, associated with both the RGC (upper valley of El Paso above American Dam) and RGR (lower valley of El Paso below American Dam) projects, is used to demonstrate how urban property values and agricultural production as well as their potential damage due to flooding are estimated. In El Paso County, residential properties in the revised levee-associated flood zone cover 1070 ha (2643 acres), commercial development 1117 ha (2759 acres) and industrial development 13 ha (32 acres), respectively, while crop land covers 953 ha (2356 acres). Agricultural land acreage was further identified by crop type using MRLC maps to obtain a more accurate estimation of crops that have different levels of damage.

Urban Property: El Paso County

Table 3 provides details on the urban structures included in the analysis (residential, commercial, and industrial) for El Paso County, using shape files from the appraisal district linking inundated properties, structure, and values. Individual and total appraised property values were determined. Using the total appraised value, the content value (i.e. 50%) for residential properties, as well as the inventory value (i.e. 73%) and equipment value (i.e. 63%) for commercial and industrial development, were determined. The appropriate flood-damage coefficients were then applied (to the value of the structure and improvements) to derive total damage for each of the classifications (i.e. contents, inventory, and equipment). In this case, the estimated damage to the structure, content, and inventory for residential properties is \$108.6 million, \$26.9 million for commercial, and \$2.4 million for industrial. Potential damage in El Paso County is estimated at \$137.9 million for urban uses.

Agriculture Uses: El Paso County

Table 4 shows details of agricultural damage estimates due to flooding for El Paso County. The individual crops are listed with expected non-flood yield, percentage yield loss due to flooding, crop price (normalized prices), revenue loss, reduced harvesting costs, and net damage per hectare. The crop types, i.e. orchard and row crops as defined using MRLC maps, orthophotos and appraisal district property parcel maps, are used to calculate total acreage of each crop. Note that cotton seed is a joint product with cotton lint. The total potential agricultural damage for El Paso County is estimated at \$1.2 million or \$1310 per hectare. This area is characterized by high value agricultural products.

Total Damage/Losses Avoided: El Paso County

As shown in Table 5, potentially flood-impacted land use in the El Paso County revised flood zone is evenly dispersed between residential, commercial, and agricultural uses across the estimated 3153 ha (7790 acres). The area most affected by a levee failure, however, is the estimated 1070 ha (2643 acres) of residences (i.e. found primarily in the City of El Paso, Texas), which would sustain an estimated 78% of the total damage. Although agricultural land use encompasses about 30% of the floodplain's land area, it is anticipated to only incur 1% of total damage. In summary, a levee failure in this project area is estimated to significantly damage approximately 4591 residential, commercial, and industrial structures and 953 ha (2356 acres) of agricultural land, resulting in a total of $139.1 million in economic damage.

Table 3. Estimated economic damage for urban land use in El Paso County attributable to a levee failure of the RGC and RGR projects, 2004

	Urban land use			
Category	Residential	Commercial	Industrial	Total
Value				
Structure				
Number of structures	4 251	339	1	4 591
Total appraised value ($, million)	264.56	52.41	4.67	321.64
Average value per structure ($)	62 236	154 577	4 670 297	70 058
Contents, inventory, equipment				
Contents value (percentage of appraised value)	50.0	—	—	—
Contents value ($, million)	132.28	—	—	132.28
Inventory value (percentage of appraised value)	—	73.0	73.0	—
Inventory value ($, million)	—	38.25	3.41	41.66
Equipment value (percentage of appraised value)	—	63.0	63.0	—
Equipment value ($, million)	—	33.01	2.94	35.95
Damage				
Structure				
Damage factor (percentage of appraised value)	32.1	11.1	11.1	—
Damage value ($, million)	84.93	5.82	0.52	91.27
Contents				
Damage factor (percentage of contents value)	17.9	—	—	—
Damage value ($, million)	23.68	—	—	23.68
Inventory				
Damage factor (percentage of inventory value)	—	36.0	36.0	—
Damage value ($, million)	—	13.77	1.23	15.00
Equipment				
Damage factor (percentage of equipment value)	—	22.0	22.0	—
Damage value ($, million)	—	7.26	0.65	7.91
Total damage for urban land use ($, million)	108.61	26.85	2.39	137.85

Aggregate Damage/Losses Avoided: All Projects

As depicted in Table 6, the estimated gross value of economic losses avoided by IBWC flood-control infrastructure on the US side of the border for the four project areas analysed totals $322.9 million for both developed property and agriculture. There is a wide range of difference in damage between project areas, with the most populated project area (and most developed) being the LRGVFC project with $167.2 million of avoided economic losses. The infrastructure along the RGC project provides protection against losses of $13.7 million for Doña Ana County in New Mexico and a portion of the $139.1 million in losses avoided for El Paso County in Texas (the upper valley portion above American Dam) as discussed in the previous section. The remaining $139.1 million in economic losses for El Paso County (below American Dam) is avoided by the RGR project. The PVFC project area is the smallest project with the least amount of development, and has an estimated $2.9 million in avoided economic losses (Table 6).

The urban land use category receives the largest amount of avoided economic losses, a total of $302.2 million, provided by IBWC flood-control infrastructure. In particular, residential structures are provided an estimated $220.9 million in avoided economic losses while commercial and industrial structures are provided $79 million and $2.4 million, respectively (Table 6). Similarly, the number of structures benefiting from levee protection is comparable across urban land use. It is estimated that 11 010 residential structures are in the revised floodplain (i.e. FEMA less 'disconnected areas'). Commercial and industrial structures affected are substantially lower with 1156 and 1 structure(s), respectively.

Agricultural land use, though less affected (in terms of dollars) than urbanized land, is provided an estimated $20.7 million in economic losses avoided by levees and other flood-control infrastructure protecting an estimated 32 880 ha (81 249 acres).

Discussions and Recommendations

The baseline gross-value estimates of avoided losses provided by flood-control infrastructure are applicable towards a complete cost–benefit analysis. With a brief study period and other limiting constraints, several broad-based assumptions were made on key issues. For example, only levees and flood inundations in the USA were considered in this analysis. The gross-value estimates are in 2004 dollars, i.e. assuming flooding occurred in 2004. Since flood inundation maps were not available for all areas, the total flood area was estimated for each project area. This entailed the use of digital copies of FEMA's 100-year flood maps and FIRMs to make extrapolations. The flood inundation depth was unknown for the hypothetical flooding, but was assumed to be 2 feet across the estimated flood inundation area. The results are point estimates (i.e. deterministic and without sensitivity analyses or stochastic elements) based on a single hypothetical flood event. As such, there was no accounting for risk and/or uncertainty about the frequency of occurrence or the actual range in the data input. The estimates were based upon a single flood event without probabilities of location, timing, severity, duration, etc. The gross-value estimates account for developed-property and agricultural crop damage only. As such, the following categories and types of loss have limited or no inclusion: loss of human life; loss of livestock; loss of transportation and transportation infrastructure, vehicles, roads and highways, bridges, and dams; disaster mitigation costs; other agricultural costs, farm equipment (e.g. tractors, underground wells, tillage implements), and loss and/or

Table 4. Estimated economic damage for agricultural land use in El Paso County attributable to a levee failure of the RGC and RGR projects, 2004

Crop	Unit	Base yield	Flood yield	Yield loss (units per hectare)	Crop price ($ per unit)	Harvest costs ($ per unit)	Revenue loss ($ per hectare)	Increased flood cost ($ per hectare)	Total damage ($ per hectare)	Reduced harvest costs ($ per hectare)	Net flood damage ($ per hectare)	Percentage acreage allocation	Specific crop acreage (ha)	Total potential damage ($1000)
Row crops														
Alfalfa	Tons	12.1	9.6	2.5	152.00	11.89	368.09	0	368.09	28.79	339.30	14	134	45
Pima cotton	Pounds	2637	0	2637	0.88	0.16	2319.20	0	2319.20	430.43	1888.77	58	553	1044
Cotton seed[a]	Pounds	4065	0	4065	0.05	0.00	203.24	0	203.24	0.00	203.24	NA[a]	NA[a]	112
Red chile	Tons	2.97	0	2.97	1375.00	0.24	4077.22	0	4077.22	0.69	4076.53	1	9.71	39
Wheat	Bushels	72	0	72	3.00	0.51	214.98	0	214.98	36.62	178.36	5	48	8.5
Other														
Pecan	Pounds	4942	4942	0	0.90	0.35	0.00	0	0.00	0.00	0.00	22	210	0
												Total	953	1248.5

NA, Not applicable.
[a] Cotton seed is a co-product of cotton production, thereby not requiring additional acreage.
Source: Texas Agricultural Statistics Service (2002).

Table 5. Estimated economic flood-control benefits in El Paso County provided by the RGC and RGR projects, by land use category and type, 2004

Land use category and type	Estimated area (ha)	Percentage of total area	Estimated damage ($ per hectare)	Number of structures	Estimated damage total ($, million)	Percentage of total damage
Urban						
Residential	1070	34	101 538	4251	108.60	78
Commercial	1117	35	24 048	339	26.85	19
Industrial	13	1	184 792	1	2.39	2
Subtotal	2199	70	62 685	4591	137.84	99
Agriculture						
Subtotal	953	30	1 310	—	1.25	1
Total	3153	100	44 123	4591	139.10	100

displacement of topsoil; and environmental impacts (wildlife and endangered species, chemical spills).

The above-mentioned limitations highlight significant cost variables not considered as determinants in the reported estimates. The reported estimates of avoided losses accruing to IBWC flood-control infrastructure should be viewed considering the limitations defined. Additional research efforts targeting one or more of the study's shortcomings would enable a more detailed and accurate estimate of benefits provided by IBWC flood-control infrastructure along the Rio Grande. Therefore, the authors recommend further development and refinement for the following topics: (1) flood inundation maps, supported by appropriate hydrological and hydraulic studies focused on levee and dam failure; (2) detailed field surveys of structures within the floodplain; (3) detailed stage-damage curves for an expanded set of urban land use types; (4) more recent satellite imagery combined with the inclusion of a broader set of urban land use types and their associated values; and (5) other economic, socio-economic, and environmental cost estimates for the specified study area.

Conclusions

Innovative integrated GIS methods were developed and applied as an effective tool for a rapid economic assessment of potential flood-control economic losses for a single event for four USIBWC flood-control projects along the Rio Grande. Several forms of maps, images, and a variety of data including multiple shape files (FEMA maps, property parcels) and imagery (DOQQs, MRLC maps, LandSat 7 images), were integrated through GISs to define flood inundation areas, identify types of land use, determine acreage for each land type, and extract information needed for economic assessment in a timely manner.

Multiple economic assessment methods were employed for urban property values and agricultural production as well as their expected damage from flooding based on the best information available and applicability of the method. Estimates of representative residential, commercial, and industrial property values and agricultural production values were developed from property assessment records, economic development councils, crop enterprise budgets and cropping patterns, census data, and previous USACE flooding

Table 6. Summary of estimated economic benefits provided by four USIBWC Rio Grande flood-control projects, 2004

	Total property damage[a]										
	Agricultural land use				Urban land use						
	composite[b]		Residential		Commercial		Industrial		Total property		
Project area	No. of hectares	Subtotal damage ($, million)	No. of structures	Subtotal damage ($, million)	No. of structures	Subtotal damage ($, million)	No. of structures	Subtotal damage ($, million)	No. of structures	Total damage ($, million)	Percentage of total damage
Doña Ana County, NM (RGC)	1 005	1.82	397	6.78	67	5.08	0	0	464	13.68	4.20
El Paso County, TX (RGC and RGR)	953	1.25	4 251	108.60	339	26.85	1	2.39	4 591	139.10	43.10
PVFC	309	0.09	589	2.84	0	0	0	0	589	2.93	0.90
LRGVFC	30 613	17.52	5 773	102.63	750	47.07	0	0	6 523	167.22	51.80
Total	32 880	20.68	11 010	220.86	1156	79.00	1	2.39	12 167	322.93	100.00
Percentage of total damage		6.40		68.40		24.50		0.70		100.00	

[a] These gross-value estimates only account for agricultural crop and developed-property damages on only the US side of the USA–Mexico border.
[b] The specific crops differ by reach and include varying proportions of row crops, vegetables, citrus, etc.

studies, etc. Gross economic values of avoided losses for a sample of each of the land use types were determined and extrapolated to similar land use areas in the flood zone. Economic losses avoided by each individual project area and an aggregate estimate of the economic losses avoided by flood protection were determined.

Economic losses avoided by developed properties and agricultural production by the RGC and RGR projects are estimated at $13.7 million (2004) for Doña Ana County in New Mexico and $139.1 million for El Paso County in Texas, respectively. The PVFC project provides $2.9 million in avoided economic losses, while the LRGVFC project provides $167.2 million. Combined, the IBWC flood-control infrastructure in four project areas provides $322.9 million in avoided economic losses, of which $302.2 million is for the developed properties and $20.7 million is for agriculture.

Given a short project time-line and the large geographic extent of the river and floodway levee system, economic estimates for four IBWC projects only related to the US side of the border. In addition, the estimate of the potential damage is limited to property damage only. This study is preliminary in nature and is expected to provide an underestimate of the total flood-control protection benefits of certain IBWC flood-control infrastructure along the Rio Grande. Several benefit variables are not incorporated into the reported baseline estimates; hence, the results need to be interpreted considering the limitations of the study.

In recognizing the above-mentioned limitations, further development and refinement are recommended for the following topics to achieve more accurate and defendable estimates: (1) flood inundation maps; (2) detailed field surveys of structures; (3) detailed stage-damage curves; (4) more recent satellite imagery; and (5) other economic, socio-economic, and environmental cost estimates.

Acknowledgements

This study was funded in part by IBWC and based on the work developed under the Cooperative State Research, Education, and Extension Service, US Department of Agriculture agreement no. 2003-34461-13278 (Rio Grande Basin Initiative on Water Conservation). Many individuals have contributed to the methodology development, data compilation, and economic analysis. The authors gratefully acknowledge and appreciate the collaboration and assistance from Jim Robinson, Albert Moehlig, Raymundo Aguirre, Gilbert Anaya, and Sherry Brinkley of IBWC, Jim Stefanov of USGS (former project manager with USIBWC), the El Paso Central Appraisal District, City of El Paso, the Presidio Appraisal Office, and the Marfa County Tax Office, respectively. Invaluable contributions to the analysis were made by Joshua Villalobos, Naomi Assadian, and Wendy Morrison at the El Paso Research Center, James T. McGuckin of New Mexico State University, Jennifer H. Jacobs and Chris Staats at SSL, and Roger Freeman. Any omission or errors are the responsibility of the authors. The authors also appreciate Dr Chennat Gopalakrishnan, the editor and an anonymous reviewer for their constructive comments.

References

Agribusiness Council of El Paso (2000) *The Importance of Agribusiness to El Paso County* (College Station, TX: TAES, Texas A&M University).

City of El Paso (2004a) Population 2002. Available at http://www.elpasotexas.gov/econdev/_documents/Population_2002.pdf (accessed December 2004).

City of El Paso (2004b) Economic overview. Available at http://www.elpasotexas.gov/econdev/eco_overview.asp (accessed December 2004).

FEMA (1985) *Flood Insurance Rate Maps* (Washington, DC: FEMA).

FEMA (2004a) Overview of HAZUS-MH (multi-hazards). Available at http://www.fema.gov/hazus/hz_meth.shtm (accessed December 2004).

FEMA (2004b) Flood hazard mapping. Available at http://www.fema.gov/fhm/fhm_srch.shtm (accessed December 2004).

Freeman, R. (2004) Personal communications [economic consultant, formerly with USACE, Galveston District].

IBWC (2003) Hydraulic model of the Rio Grande and floodways within the Lower Rio Grande Flood Control project. IBWC internal report, El Paso, TX.

IBWC (2004a) Project descriptions, origins and histories for various IBWC projects with multiple completion dates. In-house and unpublished documentation provided by USIBWC in electronic format, El Paso, TX.

IBWC (2004b) Scope of work (IBM 92-21): analysis of flood control benefits for USIBWC Rio Grande flood control projects. IBWC internal report, El Paso, TX.

International Federation of Red Cross and Red Crescent Societies (2002) World disasters report 2002. Available at http://www.ifrc.org/puplicat/wdr2002/ (accessed November 2004).

Lacewell, R. D. & Freeman, R. L. (1990) *Agricultural Benefit Estimator: Documentation* (College Station, TX/Galveston, TX: TAES/USACE).

Lacewell, R. D., Robinson, J. R. C., Stoll, J. R., Freeman, R. L., Kazda, J., Hankamer, C., Stewart, R., Onyeji, C. & Taylor, M. (1989) Estimated agricultural benefits to drainage and flood control in Cameron County, Texas. Final report prepared for US Department of Agriculture, Soil Conservation Service, Temple, TX and USACE, Galveston, TX.

Lacewell, R. D., Purvis, A., Freeman, R. L., Kazda, J., Petit, D., Ruspasinghe, A., Shepperd, J. & Taylor, M. (1995) Estimated south main channel agricultural benefits attributable to drainage and flood control in Hidalgo and Willacy counties, Texas. Final report prepared for USACE, Galveston, TX (College Station, TX: TAES).

Lacewell, R. D., Ellis, J. R., Chowdhury, M. & Freeman, R. L. (1997) *Review of the Raymondville Drain, Economic Benefits* (College Station, TX: for USACE, Galveston, TX).

Lacewell, R. D., Freeman, R. L., Petit, D. & Zinn, M. (2001) *Update of Estimated Agricultural Benefits Attributable to Drainage and Flood Control in Willacy County* (College Station, TX: for USACE, Galveston, TX).

McGuckin, J. T. (2004) Personal communications [professor, New Mexico State University, Las Cruces, NM].

Moehlig, A. (2004) Personal communications [general engineer, Technical Planning Division, IBWC, El Paso, TX].

Robinson, J. (2004a) Personal communications [former chief of Engineering Services Division, IBWC, El Paso, TX].

Robinson, J. R. C. (2004b) *Crop and Livestock Enterprise Budgets for the Lower Rio Grande Valley* (Weslaco, TX: Texas Agricultural Extension Service).

Robinson, J. R. C., Lacewell, R. D., Stoll, J. R. & Freeman, R. L. (1992) Estimating agricultural benefits from drainage over a relatively level terrain, *Agricultural Water Management*, pp. 79–91.

Shabman, L., Stephenson, K., Thunberg, E., Dietz, B., Driscoll, P. & O'Grady, K. (1998) Comparing benefit estimation techniques: residential flood hazard reduction benefits in Roanoke, Virginia. IWR Report 98-R-2, USACE, Alexandria, VA.

Stefanov, J. (2004) Personal communications. [associate director, USGS, Austin, TX].

Texas Agricultural Statistics Service (2002) Texas Agricultural Statistics 2002, Compiled for U.S. Department of Agriculture and Texas Department of Agriculture, Austin, Texas.

Texas Cooperative Extension (2003) Planning budgets. Available at http://jenann.tamu.edu/budgets/district/12/2003 (accessed December 2004).

Texas Department of Agriculture (1998) *Texas Agricultural Statistics* (Austin, TX: Texas Agricultural Statistics Service).

Texas Water Development Board (2002) *Water for Texas—2002* (Austin, TX: Texas Water Development Board).

US Bureau of Reclamation (2004) *Inundation Model for the Rio Grande Project* (Denver, CO: US Bureau of Reclamation).

US Department of Agriculture (2003) *Normalized Prices to be Used for National Economic Analysis as Outlined in the Principles and Guidelines* (Washington, DC: Economic Research Service).

US Water Resources Council (1983) *Economic Environmental Principles and Guidelines for Water and Related Land Resources Studies* (Washington, DC: US Government Printing Office).

USACE (1982) *Lower Rio Grande Basin, Texas Flood Control and Major Drainage Project* (Galveston, TX: USACE).

USACE (2003) *Economic Guidance Memorandum (EGM) 04-01: Generic Depth Damage Relationships for Residential Structures with Basements* (Washington, DC: Headquarters, Civic Work, Planning and Policy Division (CECW-PG), USACE).

USACE (2004) Galveston District historical stage-damage curves. Unpublished

Adoption of More Technically Efficient Irrigation Systems as a Drought Response

ERIC C. SCHUCK, W. MARSHALL FRASIER, ROBERT S. WEBB, LINDSEY J. ELLINGSON & WENDY J. UMBERGER

Introduction

In 2002 Colorado experienced the worst drought in the state's history. Water rights dated 1865 and 1881 placed calls on the South Platte River—the most senior calls placed on the river in over a generation—and the Rio Grande very nearly ceased to flow (Hall, 2002). The severity of this drought cannot be overstated. These conditions were rated 'exceptional' by the US Drought Monitor and were the most severe drought experienced in the region since the 'Dust Bowl' (National Oceanic Atmospheric Administration, 2002; Tronstad & Feuz, 2002). Indeed, based on studies of tree rings and archaeological evidence from aboriginal cultures, the Colorado drought was arguably the worst in recorded history (Pielke *et al.*, 2003).

Agriculture is the dominant user of water in the state, so the effects of the drought hit this sector of the economy especially hard. The state's most widely planted crop by acreage, winter wheat, saw production fall by nearly 30 million bu. (817 000 tonnes) between 2001 and 2002, a reduction of almost 60% (National Agricultural Statistics Service, 2003a). Total livestock numbers in the state fell by almost 14% from 2002 to 2003, a blow that was of great concern since livestock accounts for nearly 60% of the state's agricultural production by value (National Agricultural Statistics Service, 2003b).

While the direct economic impact of the drought is certainly of interest, an important question arises with important policy implications for irrigated agriculture: that is, to understand how irrigators responded to this drought, and what changes were made in production practices to alleviate the effects of this and future droughts.

Irrigators can respond to drought in a variety of ways. In the short term, they can reduce water applications, fallow acreage or change crops. With their eye on a longer horizon, they can also adopt different irrigation systems. This latter response is of great concern. While changes in irrigation technology can allow producers to continue production on lower levels of water applications, changes in return flows stemming from changes in irrigation technology can have profound impacts on both downstream flows and water quality. As a result, while more technically efficient irrigation systems can potentially help individual farms survive droughts, the regional effects may or may not be helpful. Using farm-level survey data from Colorado irrigators, this research assesses if irrigators in Colorado adopted new irrigation systems in response to the 2002 drought, how prevalent this response was, what factors influenced this decision and what the potential implications are for overall water supplies in the state as a result of any observed changes in irrigation technology.

Background

Adoption of more technically efficient irrigation systems may be a way to facilitate production during severe droughts by maintaining consumption with reduced applications. More technically efficient irrigation systems, such as low-pressure sprinkler or drip systems, transmit a higher proportion of applied water to the root zones of crops. This allows a higher level of water consumption for crops at a given level of water application, and can allow an irrigator to reduce irrigation application rates while still meeting the consumptive demands of the crops.

Improvements in irrigation efficiency frequently correspond to moderate yield improvements (Dinar & Zilberman, 1991) and can also reduce production costs and decrease salinization (Wichelns, 1991). Both yield effects and cost reductions can improve on-farm profits and should promote adoption. However, not all irrigators will be able to take advantage of these alternative technologies or management practices. On-farm variations in land attributes, historical cropping patterns and water costs may make changes in irrigation systems that are generally more profitable less so in specific cases (Green *et al.*, 1996; Green & Sunding, 1997; Schuck & Green, 2001).

The existing irrigation technology adoption literature is both extensive and well developed (see Caswell & Zilberman, 1985; Lichtenberg, 1989; Dinar & Yaron, 1990; Negri & Brooks, 1990; Dinar & Zilberman, 1991; Shrestha & Gopalakrishnan, 1993; Green *et al.*, 1996; Green & Sunding, 1997; Schuck & Green, 2001). In general, higher water costs have been found to lead to adoption of more technically efficient irrigation systems, as has lower-quality water. However, the existing literature on the adoption of improved irrigation technology tends to emphasize price as an indicator of scarcity rather than actual measures of physical scarcity. Additionally, most of the cited literature focuses on the unique irrigation conditions of Israel's Negev or California's San Joaquin Valley.

While these areas are important areas of production, they are not necessarily representative of other irrigated areas. In particular, their institutional settings are very different from Colorado. Specifically, while the San Joaquin Valley receives the majority of its water through the US Bureau of Reclamation (USBR) via service contracts between regional irrigation districts and irrigators, most water in Colorado is privately held and distributed through either direct diversions or private mutual share ditch companies (Frasier *et al.*, 1999). As a result, unlike California, where irrigators typically pay a volumetric price to the USBR or an intermediate water provider to obtain water, Colorado irrigators already hold the right to the water, either individually or collectively, and do not need to pay a volumetric price to acquire water. Rather, they pay a fraction of the costs of the delivery infrastructure and these costs are typically not volumetric. Consequently, while existing irrigation technology literature from California emphasizes the role of volumetric price in determining irrigation technology choices, this is not a variable that Colorado irrigators face.

What this means is existing irrigation technology adoption literature reflects an institutional structure and irrigation environment that are very different from Colorado and other parts of the western USA. Results from these regions may not be compatible with the Colorado experience, particularly during droughts. The most important difference between California and Colorado relates to water price. Since most irrigation water in Colorado is privately held, few irrigators face a discernible marginal water price. In the absence of a known marginal price, existing models of irrigation technology adoption do not accurately describe Colorado. This divergence between publicly held 'government' water and privately owned water can lead to different incentives for irrigators in Colorado than are seen in other regions. Without a marginal price to observe as a measure of scarcity during droughts, it is necessary to use physical measures of water supply in their place. This research examines how irrigators in Colorado adopted improved irrigation technology in response to reduced on-farm water supplies during the 2002 drought, extending existing irrigation technology models to better reflect the institutional environment of Colorado and other settings where irrigators do not face a price for water.

The other critical difference between Colorado and other irrigated areas relates to farm structure. Many of the irrigation technology adoption models from California describe a single crop grown under a single irrigation system (see Green *et al.*, 1996; Green & Sunding, 1997; Schuck & Green, 2001). Irrigated farms in Colorado tend to grow several crops under a variety of adjacent irrigation systems. Consequently, it is possible to observe several different types of irrigation systems all being operated simultaneously on the same farm by the same manager. During a drought, an irrigator will have to evaluate the effectiveness of the entire farm's irrigation infrastructure, choosing to upgrade some portions and downgrade others depending upon the relative cost of changes to each different system. In this context, how irrigators managed their overall mix of irrigation systems observed on the farm matters more than how drought influenced adoption rates for a given system. It is therefore necessary to model irrigation technology adoption during a drought as a proportional mix of systems across the farm, not simply as a single choice.

The combination of these two differences—one in water delivery institutions, one in the nature of the farms analysed—means that existing literature on adoption of improved irrigation systems may not accurately reflect experiences in Colorado. Given the unprecedented severity of the 2002 drought in Colorado, developing irrigation technology adoption models adapted to the state is essential in assessing the full economic impacts of the drought on the state. This research uses surveys of irrigators from the 2002 drought to examine whether or not Colorado irrigators adopted more technically efficient irrigation systems in response to this extreme drought event. It does so in a Colorado-specific context that substitutes physical measures of scarcity for water prices and evaluates the mix of irrigation systems across a single farm rather than as a single crop/field choice. This paper proceeds by first developing a theoretical model of irrigation technology adoption in Colorado, and then evaluates this model using survey data taken during the 2002 drought in Colorado.

Irrigation Technology Adoption Model

One of the more fundamental problems in crop water demand is that water use in agriculture tends to be highly inelastic (Nieswiadomy, 1988; Ogg & Gollehon, 1989). Given this, changes in water use must focus at the extensive rather than the intensive margin. One of the primary means of changing water use at the extensive margin is through adjusting the efficiency of irrigation systems. The question then is simply a matter of identifying how and why irrigators choose the irrigation systems they do.

While higher water costs are associated with adopting more technically efficient irrigation systems, more recent work by Carey & Zilberman (2002) suggests that the hurdle rates for adopting more efficient systems can be quite high and are often dependent upon major external events such as drought.

The basic irrigation technology adoption problem is well established, and most research follows the model first put forward by Caswell & Zilberman (1986). Following Caswell & Zilberman (1986), the profit-maximizing irrigator will choose the *j*th irrigation system over the competing *k*th irrigation system if the *j*th system is perceived as being more profitable than the *k*th system. If π_j are the profits under the *j*th system and π_k are the profits under the *k*th system, then the condition for the adoption of the *j*th system over the *k*th system is simply:

$$\pi_j > \pi_k \qquad (1)$$

Profits are determined by the revenues and costs from production under a given irrigation system. For each farm, a farmer will maximize profits by choosing a vector of acreage allocations (l) and water applications (AW) across m crops. Production is conditioned on the physical characteristics of the farm, α, where α is a vector of field characteristics such as soil type and slope. Production for each of the m crops is a quasi-concave function of acreage allocation and effective water application which is simply the product of water applied (AW) and the application efficiency (δ_j) of the irrigation system in place. Total crop production under the *j*th irrigation technology by the irrigator is then the function $f(\delta_j AW, l; \alpha)$. If the vector of crop prices is p, then revenues from crop production are given by $p'f(\delta_j AW, l; \alpha)$.

Production costs under the *j*th irrigation system are given by the function $C_j(AW, l)$ where all variables are as previously defined. Note that while production is determined by effective water, costs are a function of applied water. This is because water that is applied but not used by the crop must still be delivered to maintain a given level of crop consumptive demand. Consequently, one of the primary advantages of a more technically efficient irrigation system is the cost reduction associated with reducing the volume of water that is delivered but is not translated into productive yield.

In addition to the costs of actually producing under a given irrigation system, the limits of available water supplies must also be recognized. For the Colorado setting, the irrigator owns some share of a regional water supply. This ownership entitles the irrigator to an annual allotment of water, ω. At the regional level, water supplies are random and the actual realized value of ω is uncertain over time, but when supplies are not scarce the maximum amount of water the irrigator can receive is given by the upper bound of the irrigator's shares, or $\bar{\omega}$. If the expected consumptive requirements for a farm in a given year are $\Omega\ AW,\ l$, then for an irrigator to actually produce in a given year the annual allotment of water must meet or exceed the expected consumptive water requirements for the farm given the farm's existing irrigation system. For the jth irrigation system, this implies $\delta_j\omega \geq \Omega(AW, l)$. For simplicity, ω can be represented by a binary distribution where $\bar{\omega}$ is received in full or it is not. The likelihood of $\bar{\omega}$ being received is θ and the likelihood of it being inadequate is simply $(1-\theta)$. This implies the irrigator faces the annual water supply constraint of $\delta_j\theta\bar{\omega} \geq \Omega(AW, l)$ under the jth irrigation technology.

Combining revenues and costs for the irrigator with this supply constraint gives the restricted profit function under the jth irrigation system for the whole farm, or:

$$\pi(p, \delta_j, AW, l; \alpha, \theta\bar{\omega}) = \max{}_{l,AW}[p'f_j(\delta_j AW, l; \alpha) - C_j(AW, l) : \delta_j\theta\bar{\omega} \geq \Omega(AW, l)] \quad (2)$$

Returning to the idea introduced by Caswell & Zilberman (1986) that an irrigator will choose the jth irrigation system over the kth system if it is more profitable implies that the decision to adopt a more technically efficient irrigation system under our present formulation is:

$$\pi(p, \delta_j, AW, l; \alpha, \theta\bar{\omega}) \geq \pi(p, \delta_k, AW, l; \alpha, \theta\bar{\omega}) \quad (3)$$

The adoption decision implied by Equation (3) departs from existing irrigation technology adoption models in one critical aspect: it uses the irrigator's own water entitlements and the likelihood of receiving that water in place of water price. This specification is more in keeping with the institutional environment in Colorado and reflects the potential for a drought to induce a change in irrigation technology by disrupting the delivery of water.

The adoption decision suggested in Equation (3) can be transformed into an empirically estimable discrete choice model by assuming irrigators maximize their profits in a random utility framework (Ben-Akiva & Lerman, 1985). Unfortunately, while equation (3) provides a general rule for adoption of more technically efficient irrigation systems that is compatible with the institutions of Colorado, as described it allows only for a discrete choice between two systems. That is, it involves only the choice of j over k. Given that farms in Colorado typically have multiple irrigation systems on different portions of the farm, what is needed is a model allowing irrigators to simultaneously choose both j and k. Fortunately, the discrete choice model can also handle this. Instead of an absolute choice between competing alternatives, the basic discrete choice model can also show the proportion of each choice selected by the decision maker. Within the context of irrigation systems this has been done to show regional irrigation technology choices (Schaible *et al.*, 1991; Schaible & Aillery, 2003), but in this instance it is at the farm rather than regional level.

Empirical Analysis

The model suggested by Equation (3) can then be empirically estimated as a proportional limited dependent variable model with the percentage of each alternative irrigation system on the farm serving as the dependent variable and water supply as an explanatory variable.

However, there is a potential endogeneity issue inherent in the modelling of irrigation technology as a response to drought. Basically, to determine if drought causes irrigators to adopt some irrigation systems in preference to others, choosing irrigation systems changes as a drought response has to be an explanatory variable in the irrigation technology choice model. However, this choice is only made if there is a drought, so it is not an exogenous decision. The two questions are simultaneous, and must be modelled as such. This means two separate questions need to be evaluated: the observed choices of irrigation systems as a function of water supply and drought; and the decision to adopt an alternative irrigation system in response to drought.

Ideally, this would be estimated as a simultaneous system of multinomial limited dependent variable models, the first showing the proportion of each irrigation system present on a farm while the second shows whether or not the observed irrigation systems were chosen in response to drought. However, such systems are quite difficult to estimate and cannot be developed using conventional econometric programs. As an alternative, it is possible to estimate two individual equations. This would normally lead to a loss of statistical efficiency, but since the relationship between the two equations is unidirectional (the equation for the proportions of each irrigation system observed depends upon the decision to change systems due to drought, but the decision to change systems due to drought does not depend upon the irrigation systems at the time of observation), one equation depends upon the other equation but the reverse is not true. The covariance relationship between the two equations is therefore upper triangular in nature and can be estimated efficiently with two independent equations (Greene, 1993). This solves the endogeneity problem related to the decision to adopt a new irrigation system during the drought, yet still makes it possible to identify how the drought affected adoption rates across alternative irrigation systems.

Given this nature of the endogeneity problem, two separate equations are estimated here. The first examines whether or not irrigators chose to adopt new irrigation systems as a drought response, while the second examines their choices of irrigation systems as a function of drought. The first equation is a simple binary choice model, while the second is a proportional model describing what percentage of total irrigated acreage on a farm is covered by each of *n* different irrigation systems. The *n* different irrigation systems modelled here are: flood gravity, gated pipe and sprinklers. These are the three dominant types of irrigation systems found in Colorado, and account for over 99% of the irrigated acreage reported in the survey. Both equations follow existing literature and assume a logistic distribution for the choices modelled, so the first equation is estimated as a binomial logit while the second equation is estimated as a multinomial logit.

Data for the two equations come from a survey of 3496 agricultural producers throughout Colorado conducted by the Colorado State University Department of Agricultural and Resource Economics in conjunction with the Climate Diagnostic Center of the National Oceanic and Atmospheric Administration. For the survey, producers with operations covering more than 50 acres (1 acre = 0.4047 hectare) were drawn from the Colorado Agricultural Statistics Service producer database and mailed a questionnaire on 25 October 2002. A single reminder letter was mailed to survey recipients 1 week later following Salant & Dillman (1994). The overall response rate to the survey was approximately 30%.

The survey contained eight sections that asked questions about the farm's basic operations, water supplies, irrigated enterprises, dryland crops, livestock, drought responses, drought pressures and socio-economic characteristics. Most critically for this analysis, the section on drought responses included a specific question on whether or not irrigators adopted more technically efficient irrigation systems in response to the drought. A specific question identifying if the observed mix of irrigation systems on the farm was a direct result of the drought makes it possible to differentiate between the effects of chronically low water supplies (obtained from the portion of the survey dealing with available water supplies) and the specific shock of the 2002 drought. Having irrigators identify if their choice of irrigation systems was a specific response to the 2002 drought raises the previously mentioned endogeneity issue.

The 1100 usable responses for the survey represented a diverse cross-section of Colorado agriculture including livestock and dryland producers in addition to irrigators. Among irrigated operations, there was also significant variation regarding water source. To evaluate the question of drought impact on irrigation technology adoption, the sample was restricted to only irrigators who received water solely from mutual ditch companies. Mutual ditch companies are the most common type of irrigation water supplier in the state and provide good feedback regarding water deliveries over time. Inability to account for blending of other water sources, particularly groundwater, precludes the inclusion of respondents with additional water sources for the empirical estimation. This restriction resulted in 231 usable observations for this analysis, 20% of all questionnaires returned in the survey and nearly half of all irrigators.

The data used in the two logit models are described in Table 1. The first section contains variables that characterize the farm resource base. The relative measure of water deliveries as a fraction of full water supply filled for the irrigator during the 2002 drought is a strong indicator of the severity of the drought. On average, respondents received less than 35% of a full water supply in 2002 with some receiving full supply and some receiving none. The land resource was characterized by producers who averaged just over 150 acres of irrigated land with an average of just under 35 acres of corn and about 65 acres of alfalfa.

Table 1. Summary of response for 2002 drought survey

	Units	Average	Minimum	Maximum
Proportion of full water supply in 2002	%	34.7	0	100
Land				
Total irrigated land	Acres	156.6	0	11 620
Irrigated corn	Acres	33.6	0	1 460
Irrigated hay	Acres	63.5	0	1 400
Land leased	%	36.7	0	100
Land under irrigation system type				
Flood or siphon tube	%	51.2	0	100
Gated pipe	%	17.1	0	100
Sprinkler	%	31.7	0	100
Modified irrigation system in response to drought		11.3		
Proportion of income from agriculture	%	47.9	0	100
Education				
High school		32.8		
Some college		23.3		
Vocational/technical degree		8.1		
Bachelor's degree		23.9		
Graduate/professional degree		11.9		
Number of observations		231		

Irrigation system type was expected to influence the rate of technology adoption. On average, respondents were irrigating just over half of their irrigated acreage using flood or siphon tube technology. About one-sixth of the acreage was serviced by gated pipe and just under one-third by sprinkler. Individual producers varied from all to none of any given technology. When asked if adjustment in their technology choice for 2002 in was in direct response to the drought, over 11% of respondents indicated that their choice was dictated by the drought.

Two important demographic variables are also considered. First, the proportion of income derived from agriculture is expected to be important. On average, respondents indicated that just under half of their income was derived from agriculture. Finally, education often plays an important role in technology adoption decisions. The level of education was represented as an integer value with: high school = 1; some college = 2; vocational/technical degree = 3; bachelor's degree = 4; and graduate/professional degree = 5.

Results

Estimation of the proportional multinomial logit model was carried out using LIMDEP (Greene, 2002). The coefficients and *t*-values for the regression estimating the proportion of farmland serviced by gated pipe and sprinkler systems, respectively, are reported in Table 2. Based on standard measures of goodness of fit for logistic regression, the regression both fits the data well and is statistically significant.

Table 2. Multinomial logit results for proportion of land by irrigation system

		Gated pipe		Sprinklers	
Variable	Units	Coefficient	*t*-ratio	Coefficient	*t*-ratio
Constant		−0.68743	−1.408*	−1.63509	−3.165****
Share percentage	%	0.00914	1.574*	−0.00190	−0.314
Total irrigated cropland	Acres	0.00045	0.550	0.00108	1.798**
Irrigated corn	Acres	0.00246	1.131	0.00315	1.658**
Irrigated land leased	%	0.00166	0.370	−0.01267	−2.555***
Changed system in 2002	0/1	0.76761	1.365*	0.59438	1.034
Level of eduction	1/2/3/4/5	−0.15635	−1.250	0.21901	1.859**
Income from agriculture	%	−0.01389	−2.427***	0.00413	0.799
Model statistics					
Unrestricted log likelihood function		−208.622			
Restricted log likelihood function		−228.382			
Likelihood ratio statistic		39.521****			
		14 degrees of freedom			
Number of observations		231			

Note: Significant at α = *0.15, **0.10, ***0.05, ****0.01.

Evaluation of individual coefficients shows that the proportion of land serviced by gated pipe is most significantly influenced by the proportion of income derived from agriculture. This is an inverse relationship where those producers with higher proportions of income from agriculture used significantly lower proportions of gated pipe. Competition for time would appear to be the driving factor in this result.

The decision to change irrigation technology in response to drought and the proportion of full water delivery received in 2002 also had a statistically significant impact on the proportion of gated pipe used. If an individual changed irrigation technology in the face of drought, they were more likely to increase the proportion of land serviced by gated pipe than those for whom drought did not prompt a change. The choice of gate pipe was further influenced by the degree of shortfall. A lesser shortfall in 2002 indicated a greater proportion of gated pipe whereas greater shortfalls indicated a shift away from gated pipe.

The prevalence of sprinkler technology was most significantly affected by land tenure, cropland acreage and level of education. The model reveals that greater proportions of land rented led to a lower proportion of sprinkler irrigation. This relationship should not be surprising as landowners' incentive to invest in capital equipment is diminished if the benefits are to be shared with a tenant. Under these circumstances, with the decision to invest in long-term irrigation technology in the hands of a landlord, one would expect a lesser rate of sprinkler adoption in land leasing situations.

Scale and intensity of land utilization had the opposite effect on sprinkler adoption. As total irrigated acreage increased, the proportion of land under sprinkler systems also increased. When scale of operation exerts pressure on labour resources, adoption of sprinkler systems is one means to alleviate the pressure. Similarly, more intensive and time-sensitive crops stand to garner more benefit from sprinkler technology. Greater acreage of corn, requiring more intensive management and being the major alternative to hay crops in the sample, translated into greater adoption of sprinkler systems.

Finally, level of education had a significantly positive effect on the adoption of sprinkler irrigation systems. The greater level of education attained the greater the proportion of irrigated land serviced by sprinkler irrigation. A number of aspects of education would be consistent with this finding such as greater technical skill required to operate most sprinkler systems or differences in lifestyle preferences.

The model can be used to estimate producer response under controlled conditions. One of the more interesting relationships revealed by the model is the proportion of land dedicated to production under different intensities of water supply. Figures 1, 2 and 3 demonstrate the expected response to water supply for gated pipe, sprinkler and flood technologies, respectively. Further, each figure differentiates the response for those who indicated that the drought prompted technology changes from those who did not.

The proportion of land dedicated to gated pipe increases as the water supply increases, as shown in Figure 1. Among those who were not prompted to change irrigation technology in the face of drought, a full water supply is predicted to correspond to an average of nearly 30% of acreage dedicated to gated pipe. For those making changes in response to the drought, the average proportion would be just over 40%. If water deliveries were only 50% of full supply, acreage proportion under gated

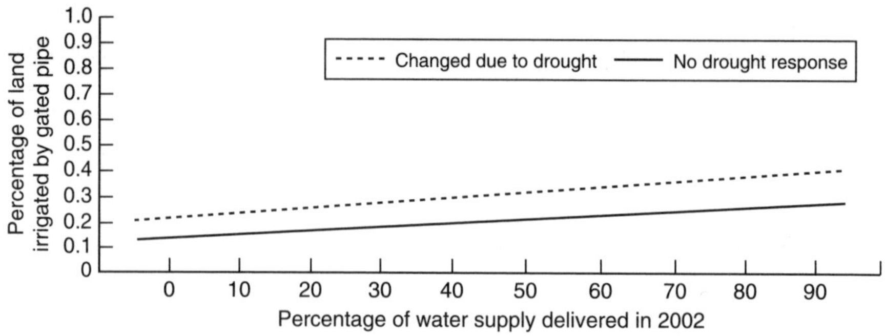

Figure 1. Proportion of land irrigated by gated pipe in response to water supply.

pipe dropped to 30% and 20%, respectively, for those responding to drought and those who did not.

As shown in Figure 2, the proportion of land under sprinkler technology followed a similar, but more intense pattern. At full water supply, land under sprinkler technology was expected to account for about 50% of all irrigated land for those not prompted to change technology and over 60% for those changing. With only half of full supplies, the proportions dropped to 46% and 36%, respectively.

Finally, the residual impact on flood irrigation technology is shown in Figure 3. Here, the impacts are the opposite of the previous two cases. As water deliveries became more available, a lesser proportion of land was serviced by flood irrigation. Further, those responding to drought through technology adoption had an overall lower rate of flood irrigation use. At full water supply, those not responding to drought would average about 20% flood, whereas those reacting by upgrading technology would have none. At 50% water supply, the proportion of land under flood increases to 44% and 24%, respectively.

What these results seem to suggest is that the drought triggered some changes in irrigation systems, but most of the movement was from gravity/siphon to gated pipe. It is worth noting that this is the least capital-intensive transition and does have a noticeable effect on irrigation efficiency. Basically, producers were making the smallest investment

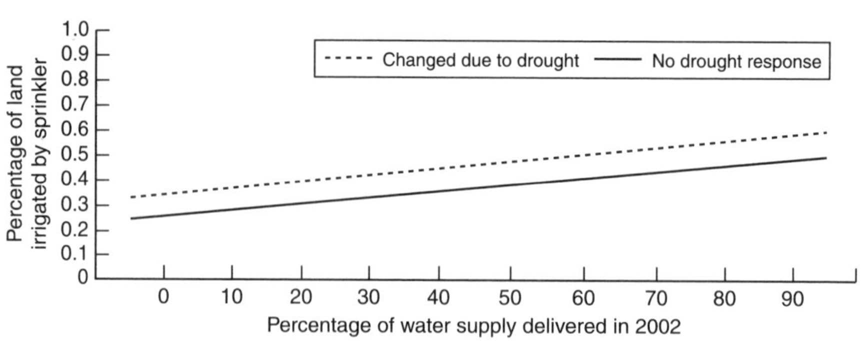

Figure 2. Proportion of land irrigated by sprinklers in response to water supply. *Note:* Evaluated at mean for total irrigated acreage, corn acreage, proportion farmland leased, education and percentage income from agriculture.

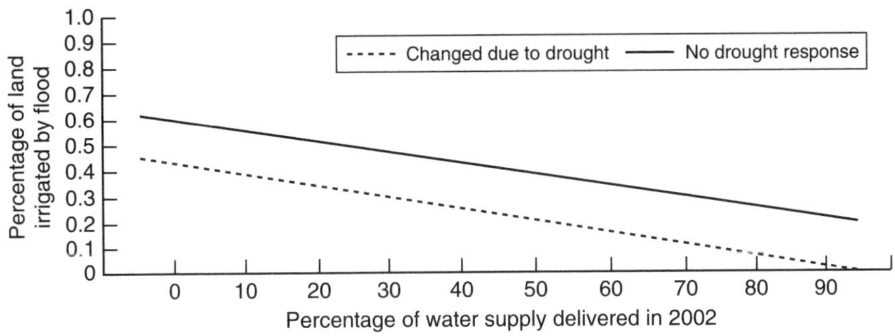

Figure 3. Proportion of land irrigated by flood in response to water supply. *Note:* Evaluated at mean for total irrigated acreage, corn acreage, proportion farmland leased, education and percentage income from agriculture.

possible to stretch their available water. In the long run, the trend might be different. In the case of this study, producers appear to have treated this as a single year disaster and not as a permanent shift in water supplies.

Conclusions

Improvements in the physical efficiency of irrigation systems are often suggested as a means of coping with short water supplies. Indeed, improvements in the technical efficiency of irrigation systems are a primary means of reducing on-farm water applications. However, little existing literature on the adoption of alternative irrigation systems focuses on response during drought periods. It is not clear how producers would respond in the face of severe drought conditions.

Econometric results reveal that source of income, scale of operation and land tenure are most significant in explaining the choice to move to more capital-intensive and, hence, more technically efficient irrigation systems. This is consistent with trends observed with the commensurate intensification that accompanies specialization in modern commercial agriculture in general. While drought itself appears to have triggered some changes in irrigation, it did not bring about wholesale changes in technology. In fact, among those who did adopt more efficient technology in response to drought, the changes tended to be relatively low-cost (e.g. moving from flood to gated pipe), suggesting a short-run response.

Acknowledgements

This research was supported by a grant from the Colorado State Agricultural Experiment Station, the National Oceanic and Atmospheric Administration and the Colorado Department of Agriculture.

References

Ben-Akiva, M. & Lerman, S. (1985) *Discrete Choice Analysis: Theory and Application to Travel Demand* (Cambridge, MA: MIT Press).

Carey, J. M. & Zilberman, D. (2002) A model of investment under uncertainty: modern irrigation technology and emerging markets in water, *American Journal of Agricultural Economics*, 84, pp. 171–183.

Caswell, M. & Zilberman, D. (1985) The choices of irrigation technologies in California, *American Journal of Agricultural Economics*, 67, pp. 223–234.

Caswell, M. & Zilberman, D. (1986) The effects of well depth and land quality on the choice of irrigation technology, *American Journal of Agricultural Economics*, 68, pp. 798–811.

Dinar, A. & Yaron, D. (1990) Influence of quality and scarcity of inputs on the adoption of modern irrigation technologies, *Western Journal of Agricultural Economics*, 15, pp. 224–233.

Dinar, A. & Zilberman, D. (1991) Effects of input quality and environmental conditions on selection of irrigation technologies, in: A. Dinar & D. Zilberman (Eds) *The Economics and Management of Water and Drainage in Agriculture*, pp. 229–250 (New York: Kluwer).

Frasier, W. M., Waskom, R. M., Hoag, D. L. & Bauder, T. A. (1999) Irrigation management in Colorado: survey data and findings. Technical Report TR99-5, Colorado State University Agricultural Experiment Station, Fort Collins, CO.

Green, G. & Sunding, D. (1997) Land allocation, soil quality, and the demand for irrigation technology, *Journal of Agricultural and Resource Economics*, 22, pp. 367–375.

Green, G., Sunding, D., Zilberman, D. & Parker, D. (1996) Explaining irrigation technology choices: a microparameter approach, *American Journal of Agricultural Economics*, 78, pp. 1064–1072.

Greene, W. H. (1993) *Econometric Analysis*, 2nd edn (Englewood Cliffs, NJ: Prentice Hall).

Greene, W. H. (2002) *LIMDEP Version 8.0, Econometric Modelling Guide* (Plainview, NY: Econometric Software Inc.).

Hall, J. (2002) Sources of water for irrigation along the South Platte during a drought, *Streamlines: Quarterly Newsletter of the Office of the State Engineer*, 16, pp. 2–3.

Lichtenberg, E. (1989) Land quality, irrigation development, and cropping patterns in the northern High Plains, *American Journal of Agricultural Economics*, 71, pp. 187–194.

National Agricultural Statistics Service (2003a) *Crop Production 2002 Summary (Cr Pr 2-01)* (Washington, DC: US Department of Agriculture).

National Agricultural Statistics Service (2003b) *Cattle Outlook on January 1, 2003* (Washington, DC: US Department of Agriculture).

National Oceanic Atmospheric Administration (2002) Climate of 2002—August US regional drought watch, National Climatic Data Center. Available at http://lwf.ncdc.noaa.govncdc.html, accessed October 2002.

Negri, D. & Brooks, D. (1990) Determinants of irrigation technology choice, *Western Journal of Agricultural Economics*, 15, pp. 213–223.

Nieswiadomy, M. (1988) Input substitution in irrigated agriculture in the High Plains of Texas, 1970– 80, *Western Journal of Agricultural Economics*, 13, pp. 63–70.

Pielke, Sr, R. A., Doesken, N. & Bliss, O. (2003) Climate of Colorado. Climatology Report 60, Department of Atmospheric Science, Colorado State University, Fort Collins, CO.

Ogg, C. W. & Gollehon, N. R. (1989) Western irrigation response to pumping costs: a water demand analysis using climatic regions, *Water Resources Research*, 25, pp. 767–773.

Salant, P. & Dillman, D. (1994) *How to Conduct Your Own Survey* (New York: Wiley).

Schaible, G. D. & Aillery, M. P. (2003) Irrigation technology transitions in the mid-plains states, *International Journal of Water Resources Development*, 19, pp. 67–88.

Schaible, G. D., Kim, C. S. & Whittlesey, N. K. (1991) Water conservation potential from irrigation technology transitions in the Pacific northwest, *Western Journal of Agricultural Economics*, 16, pp. 194–206.

Schuck, E. C. & Green, G. P. (2001) Field attributes, water pricing, and irrigation technology adoption, *Journal of Soil and Water Conservation*, 56, pp. 293–298.

Shrestha, R. & Gopalakrishnan, C. (1993) Adoption and diffusion of drip irrigation technology: an econometric analysis, *Economic Development and Cultural Change*, 41, pp. 407–418.

Tronstad, R. & Feuz, D. (2002) Impacts of the 2002 drought on western ranches and public land policies, *Western Economics Forum*, 1, pp. 19–24.

Wichelns, D. (1991) Increasing block-rate prices for irrigation water motivates drain water reduction, in: A. Dinar & D. Zilberman (Eds) *The Economics and Management of Water and Drainage in Agriculture*, pp. 275–294 (New York: Kluwer).

Rainfall Insurance: A Promising Tool for Drought Management

KIMBERLY A. ZEULI & JERRY R. SKEES

Introduction

In areas of the world where water is in short supply or where demand for water is constantly threatening to outstrip supply, diverse and complex water systems have been created to manage the risk of drought. In quasi-governmental water markets, a common institutional arrangement, a local water authority typically allocates water rights annually among competing water users based on estimates of current water reserves, future inflows, and future demand. The challenge is that water supply and demand are both highly variable across time, making accurate predictions almost impossible, not to mention extremely costly. As succinctly stated by Howitt *et al.* (1979, p. 1), "A 'normal' water year hardly exists except as a statistical average. Actually, most years are dryer than normal with the average being raised by a few extremely wet years". This process is made more difficult in the presence of drought conditions, especially since long droughts are often frequent events (Dudley, 1988).

Water supplies are inherently uncertain since they are a function of largely unpredictable weather patterns and events. Water reserves in basins and dams are depleted during dry seasons (times of high water use) and subsequently replenished during the rainy season, with important additions from snow run-off in some areas. Although the consequence of this cycle for water reserve levels can in some regions be fairly predictable over the course of a year (although this is often not the case), *El Niño* events remind the world that weather patterns and, thus, water levels are never completely predictable.

The presence of asymmetric information makes water demand estimation equally problematic. Some water users may have saved or 'banked' their previous year's water allocations or, in the case of farmers, increased their irrigation efficiency (thus requiring less water) without reporting these changes to the water authority. Water users may also have on-farm storage systems (that allow them to save any excess water) and conjunctive water systems, groundwater, and other unregulated water sources.

As a result of water supply and demand uncertainty, water authorities typically make extremely conservative annual water supply and allocation estimates. As a result, in some years they may have excess water supplies in reserve at the end of the year. Although conservative estimation certainly helps mitigate some of the risk associated with drought conditions, it is very inefficient. Stored water (whether on the farm or in a reservoir) is an imperfect insurance medium since it often leads to wasted water (Dudley, 1988). It steadily evaporates and can also lead to reservoir spills in times of excess rainfall (and/or low demand). Further, it can be assumed that water users can better optimize their temporal allocations of water since they hold more information than the water authority. The water authority may allocate more or less in a given year than will be used. The water users may also have different risk preferences than the water authority. For example, if the water authority is more risk averse than even some of the water users, it will allocate less water than will be used. If the water authority is less risk averse, water users may attempt to self-insure against drought by hoarding water rights or storing their own water.

Water markets that allow temporal trading of water allocations address some of this inefficiency. However, water markets are still inherently uncertain and the presence of uncertainty impedes the efficient functioning of any market (Arrow, 1964). The purpose of this paper is to describe how a rainfall index contract (RIC) (a relatively new insurance product) works and how it can be used to increase the efficiency of drought management in a water market context. Although several papers describe weather-based derivatives (and their potential applications in agriculture) from a theoretical perspective (e.g. Martin *et al.*, 2001; Turvey, 2001), their technicality precludes a broad readership, especially those unfamiliar with insurance literature. In addition, none has yet applied the concept to a water market. The subsequent section describes RICs in more detail for a more general audience. An established water market in New South Wales (NSW), Australia, is used to illustrate the implications of incorporating RICs into a water market. Using time series rainfall data, numerical simulations examine the impact on a hypothetical irrigator. This is followed by a more general discussion of the efficiency gains at the water market level and the implications for water demand. NSW serves as an appropriate example since it encompasses the largest irrigation industry in Australia and struggles with chronic water shortages (Crase, 1999). The paper concludes with a discussion of the challenges associated with implementing RICs and identifies future research issues.

Insurance Alternatives for Managing Water Supply Risk

All insurance products involve different types of contingent claims contracts, where a payment (indemnity) is contingent upon certain conditions or events. Insurance contracts (as opposed to futures contracts) are generally the optimal choice when the risks being insured are independent (e.g. automobile accidents). With independent risk, only a few contract holders are likely to suffer a loss at any one time. When risk is systemic, i.e. the risk of a particular event is highly correlated across individual agents, using insurance to manage risk is costly (Jaffee & Russell, 1997). For example, farmers in the same region can face correlated crop loss given adverse weather conditions such as drought conditions or excessive rain. Therefore, the likelihood that an insurance provider will have to pay a large number of claims increases as more contract holders are added to an insurance pool. As a result, private insurance companies have been unwilling to insure systemic risk.

New innovations in insurance markets are changing that norm. Private insurance and financial companies are beginning to offer several different types of contingent claim instruments, including weather-based index contracts, to cover systemic risk exposure and as portfolio diversification strategies (Skees & Barnett, 1999; Skees, 2000). The idea of basing indemnity payments on weather, such as rainfall or temperature, is not new (e.g. Sanderson, 1943). The difficulties of providing rainfall insurance in Australia spurred a debate in the 1980s (Bardsley *et al.*, 1984; Quiggen, 1986; Patrick, 1988). As Mahul (2001) points out, the primary challenge to early weather insurance tools arose from indemnity pay-offs being tied to yield effects rather than strictly to the event itself (the cause of the risk).

Insuring an outcome that is a function (even partially) of human influence gives rise to serious adverse selection and moral hazard problems. Controlling these problems (monitoring) can be costly, leading to higher insurance premiums (Miranda, 1991; Skees & Barnett, 1999; Skees, 2000). For example, water markets could introduce insurance contracts based on reservoir water levels. However, if management of the reservoir influences water levels, the water authority could manipulate the water system in a self-serving manner or manage it in such a way that would increase water supply risk (a moral hazard problem).[1] In addition, the contract might be purchased by water authorities facing the greatest water supply risk (adverse selection) and instead of improving their management or the system they would simply buy insurance. In contrast, contracts based on weather events decrease information asymmetries as well as the influence of contract agents, therefore diminishing adverse selection and moral hazard problems (making the insurance more cost-effective and increasing the likelihood that a private insurance provider would write such contracts). RICs, for instance, are based on aggregate rainfall levels (from as many years as possible) measured by reliable weather stations located in the relevant watershed. This information would be available to both the insurer and the contract holder. Further, the pay-offs under the current class of weather-based index instruments are triggered by specific and objective weather outcomes. For example, with an RIC, an indemnity payment is made to the contract holder whenever the actual rainfall measure falls below a certain percentage of the average rainfall.

RICs may not be appropriate for all water markets. Historic rainfall levels need to be highly (and positively) correlated with historic reservoir water levels. In some water systems, water levels may be more a function of temperature, snowmelt, or other water sources. In these instances, more comprehensive weather-based indexes (e.g. incorporating snow levels, temperature, etc.) would be more appropriate. If water levels are strictly a function of weather events and not management (i.e. weather events and water levels are perfectly correlated), water supply risk can also be managed with a straightforward water level contract.

Water Market Inefficiency: The Murrumbidgee Irrigation Area

The water allocation system in the Murrumbidgee Irrigation Area (MIA), an irrigation district in NSW (see Figure 1), nicely illustrates the general inefficiencies caused by water supply uncertainty and the potential for RICs. A mixed system of markets and government administration allocates water resources in the MIA (Topp & McClintock, 1998; Danzi, 1999). Since the following is a general description of the system as it was operating in 2005, it may not be a completely accurate representation of the system that currently exists for all irrigators in the MIA. Regardless, the exact details of the system are not crucial for the purpose of this paper, which is to illustrate the incorporation of insurance into a fairly common type of water market.[2]

Water entitlements are required to use water from all regulated and unregulated rivers as well as for groundwater extractions. The individual irrigator decides how many water entitlements to purchase. Although the entitlements give them the right to use a certain quantity of water, they may not receive their full entitlement. The local water authority (Murrumbidgee Irrigation) determines the annual allocations—how much they will actually receive per entitlement. There are two types of entitlements—high security and general security (Topp & McClintock, 1998). Holders of high security entitlements are guaranteed to receive their full entitlement except during severe droughts. Holders of general security entitlements, which constitute the bulk of the entitlements in NSW (Crase, 1999), have no guarantees, and usually face attenuated rights (i.e. a proportion of their entitlement). For example, an irrigator may have purchased entitlements to 100 Ml of water. If they are high security entitlements he or she may receive all 100 Ml; if they are general security he or she may receive only 5–10 Ml. Research shows that high security water allocations remain unchanged in 97 out of 100 simulated years (Danzi, 1999). In contrast, holders of general security entitlements may never receive their full entitlement (Musgrave, 1997).[3]

At the beginning of each irrigation season (July), the local water authority announces the quantity of water that will be initially allocated to entitlement (water right) holders (Musgrave, 1997; Topp & McClintock, 1998).[4] The annual water allocation decision is based on current water reserve levels and conservative estimates of future water demand

Figure 1. The MIA in NSW, Australia.

and supply (Beare *et al.*, 1998). Water demand, which varies by season and across years, is difficult to predict. Demand estimation is made more challenging by the fact that most farms in NSW have on-farm water storage systems, thus allowing them to store water for future use (Dudley & Scott, 1993). They may also use conjunctive water sources or banked water allocations from previous years (Beare *et al.*, 1998). The local water authority can (and typically does) adjust water allocations throughout the irrigation year to account for actual inflows and demand (Crase, 1999).

Estimating water supply is equally challenging, since water supply in NSW is "inherently unreliable" (Crase, 1999, p.20). In the MIA, the majority of the reserved water is contained in two water reservoirs, the Burrinjuck and Blowering dams (Beare *et al.*, 1998). The Burrinjuck Dam has highly variable inflows in winter and early spring while the Blowering Dam is managed with regulated flows from a hydroelectric dam upstream. The dams generally start to be replenished in May and by early December the cumulated inflows into the reservoirs will largely determine the water supply available for irrigation during the summer growing season (planting occurs in October–November with harvest in February). To illustrate water supply volatility, the average monthly water level index (representing an average seasonal cycle based on water level data from 1968 to 1998) is compared to the actual monthly water level index for Blowering Dam in Figure 2. Several years clearly exhibit significant shortfalls: 1968, 1982, and 1997. Water levels also fell well below normal in 1979, 1980, 1987, and 1994.

Irrigators may purchase water entitlements from the water authority, or may trade water rights with other users, at any time during the irrigation (or 'water') year (July–June typically).[5] Thus, they make an initial *ex ante* decision about how many entitlements to purchase (or retain) at the beginning of the irrigation year, after the annual allocations are announced but well in advance of the summer planting season (October–November).[6]

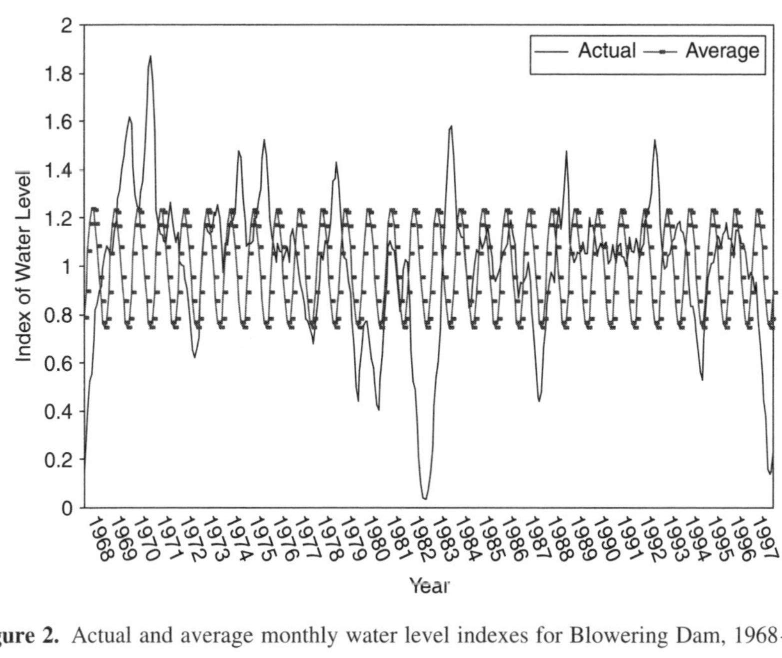

Figure 2. Actual and average monthly water level indexes for Blowering Dam, 1968–98.

When making this decision, and the related decision about how many acres to plant, the irrigator faces four sources of uncertainty. First, it is difficult to predict water demand for an entire season given the inherent uncertainty of weather and crop health. Second, even if the irrigator was able to make a perfect prediction, the water authority can change its annual allocations throughout the season, providing more or less water than anticipated. Third, the price of water rights changes throughout the season, rising as demand outstrips supply and vice versa. Fourth, the crop price is also uncertain (unless the irrigator is able to lock in a contract price).

Water entitlement and planting decisions based on incorrect expectations can be costly. For example, the underestimation of water requirements or the overestimation of intra-season reallocations could leave farmers with either severe yield loss or increased production costs since water prices would most likely rise. In an optimal situation, high water prices would be offset by high commodity prices (due to a commodity supply shortfall).[7] However, this would require at least some (high value) irrigators being able to purchase adequate water supplies to save their crops from others (low value users). In response, risk averse irrigators often self-insure (purchase more entitlements than they expect to use) to mitigate these risks. A 1996–97 survey of irrigators in NSW indicated that many hold excess entitlements as part of a hedging strategy (Beare *et al.*, 1998). In addition to decreasing the efficiency of the water market, hoarding water rights is also potentially costly for the irrigator. Although they can trade their entitlements, in seasons with sufficient rainfall or high allocations (i.e. when they are not needed) there would probably be low demand and low prices. Further, potential bumper crops could drive down crop prices and the irrigators' revenues might not ultimately cover their production costs (which include the water rights). The following section looks at the potential impact an RIC could have on managing the risks associated with estimating water demand and supply.

Constructing an RIC for the MIA

If water supply in the MIA was strictly a function of weather events, the most straightforward insurance contract for MIA irrigators would be tied to reservoir water levels. Since management of the water system (via the dams) does, however, influence the water supply, this type of insurance instrument is not appropriate (for the reasons elucidated in the second section, i.e. insuring management decisions is not efficient). If rainfall and water levels in the two dams are correlated, a better alternative would be an insurance contract based on rainfall data. To estimate this correlation for the MIA, the authors obtained monthly rainfall data over 110 years (1889–1998) from three rainfall stations in the reservoirs' watershed as well as monthly water levels for Blowering and Burrinjuck dams over 31 and 13 years, respectively.

Because the monthly water levels in the two dams were 88% correlated (a Pearson correlation coefficient) during the 13 year period their data sets overlapped, and the time series data for Blowering Dam were lengthier, only Blowering Dam data were used for the regression estimation. Using 31 years (1968–98) of matching data, a simple linear regression with annual Blowering Dam December water levels as the dependent variable and the annual cumulative July–November rainfall as the explanatory variable was estimated.[8] The model had an R^2 of 70%, meaning rainfall during July–November explains over 70% of the variation in Blowering Dam water levels for the month of December. This finding suggests that it would be reasonable to use an RIC to manage the weather-related water supply risk in the MIA. With comprehensive time series data on all weather events that impact water levels in the two dams (e.g. snowfall, temperature, etc.), a more accurate weather-based index could be created. The authors were unable to obtain the data required to test these ideas.

The following section illustrates how an RIC would be constructed. The contract would be based on average cumulative rainfall levels for July–November. For each month i (i = July, August, …, November) in each year t ($t = 1, 2, \ldots, T$; $T = 31$), the rainfall levels reported from the three weather stations are averaged and this is used as the actual rainfall level for that month and year (R_t^i). The cumulative rainfall for July–November for each year t requires summing the actual monthly rainfall levels ($R_t^c = \sum R_t^i$). The observed frequencies of cumulative rainfall during these 5 months (represented by the bars) as well as a rainfall probability distribution estimated from these observations (assuming a gamma distribution) are shown in Figure 3.

A trigger rainfall level (TR) (the rainfall level that 'triggers' an insurance payment) for any time period t is based on the average cumulative rainfall (over 31 years) and the insurance coverage level (C) purchased by the contract holder for that year:

$$TR_t = \left(\sum_t R_t^c / T\right) * C. \tag{1}$$

The coverage level is a percentage chosen by the contract holder (an alternative way to think about the coverage level is that C = 1-the deductible). For example, a 70% coverage level indemnifies the contract holder in all years when actual rainfall levels fall below 70% of the average rainfall. In years when rainfall is less then the average, but greater than 70% of the average, the contract holder would not receive any indemnity. Clearly, lower coverage levels would only insure the holder against infrequent (low probability) events. To illustrate this point, assume an average rainfall of 100 cm. In most regions it is more likely that the insured would experience rainfall levels that were less than 70% of the average (less than 70 cm) than rainfall levels that were less than 20% of the average

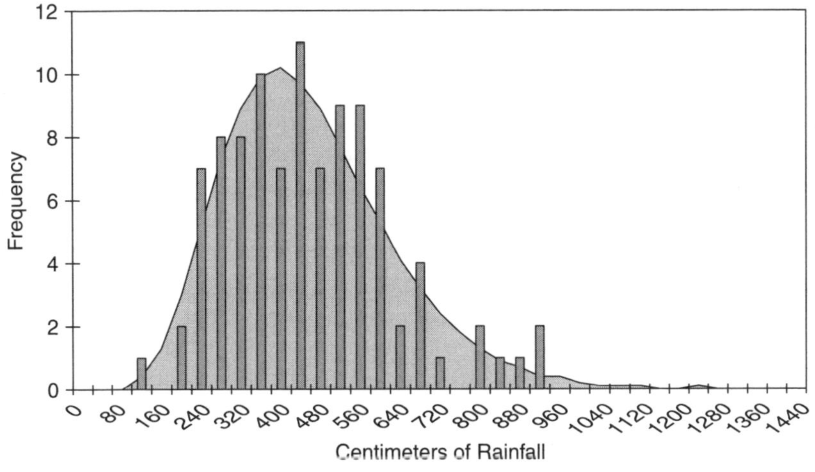

Figure 3. Probability distribution of cumulative rainfall during July–November 1968–98.

(less than 20 cm). In other words, in most regions drought conditions are less frequent than below normal rainfall levels.

The amount of money actually paid to the contract holder, the indemnity payment (P_t), is a function of the differential between the actual cumulative rainfall and trigger rainfall levels as well as the level of protection purchased by the contract holder for that period (D_t):

$$P_t = [(TR_t - R_t^c)/TR_t]^* D_t. \tag{2}$$

The level of protection, or liability, is the value the contract holder chooses to compensate himself for the anticipated loss (e.g. the value of a car if it cannot be salvaged or, in this case, the value of lost yield when a crop cannot be irrigated). For ease of calculation, assume the contract holder chooses a \$1 million liability. If the trigger rainfall level was 100 cm and the actual cumulative rainfall was 50 cm, then the payment to the contract holder is \$500 000.

Since an RIC is a contingent claim contract that serves as a risk-hedging mechanism, optimal hedging rules apply when choosing a liability level (D). Therefore, it may be optimal for the irrigator to purchase more liability than the current value of the assets at risk. For instance, as discussed above, when there is a water shortage, water prices tend to increase. If a farmer purchased a rainfall contract to hedge against both drought conditions (lower crop yields and revenue) and higher water prices (i.e. water supply risk and water price risk), he or she would take the number of units of water they normally need and multiply that by some higher than average water price to determine how much of their income is at risk and thus how much liability to purchase. However, as this RIC is specified, the indemnity payment is also affected by the chosen coverage level (C). A higher coverage level also increases the payment since it leads to a higher trigger level, which in turn means a higher percentage of the liability is paid. Higher coverage levels also increase the probability that the contract holder will receive an indemnity payment. The choice of coverage and liability levels also affects the price (premium) of the RIC. As with any insurance product, the premium is a function of the frequency and severity of the event. Thus, higher coverage levels or higher liability levels both increase indemnity payments and hence contract price.

RIC Implications for Irrigators and Water Authorities

There are several possible ways to introduce RICs into a water market. In the most straightforward approach, the water authority could simply sell rainfall contracts to water right holders any time prior to the indemnified rainfall season (July–November). Thus, the irrigator would know the current reservoir level but would be uncertain about rainfall levels during the key months that predict water supply for the summer growing/irrigation season (October–February). They would also be uncertain about the indemnity payments, which they would receive at the end of the planting season when actual rainfall measures were collected. The farmers who purchase an RIC would be insured against attenuated water allocations (i.e. insufficient irrigation or costly water) during the summer growing season.

The indemnity payments that would have been made to a hypothetical contract holder in the MIA, under various coverage levels, during 1968–98, are reported in Table 1. Liability was assumed constant at \$1 million. The actuarially fair premiums for 100%, 90%, 80%,

70%, and 60% coverage levels are, respectively, \$92 600, \$67 700, \$47 400, \$34 800, and \$24 500.[9] The cumulative rainfall index involves dividing the cumulative rainfall for each year t by the average annual cumulative rainfall for all years (1968–98), as specified in the following equation:

$$I_t = R_t^c \Big/ \left(\sum_t R_t^c / T \right). \tag{3}$$

Thus, the index number provides the percentage of the average annual cumulative rainfall received in any year t. For example, $I = 0.21$ in 1982 means that only 21% of normal (average) rainfall was accumulated during July–November. Since the average cumulative rainfall was 438 cm, this means only 91.98 cm of rainfall actually fell in 1982.

At the highest coverage level (100%), indemnity payments were triggered in 10 of the 31 years; these are the same 10 years of Blowering Dam water level shortfalls apparent in Figure 2. The contract holder would have received the highest payment in 1982. Substantial indemnity payments would also have been made in 1997 and 1980, across all coverage levels. At the lowest coverage level (60%), the contract holder would only have received payments in 3 of the 10 years. As with any insurance instrument, the potential exists for the RIC holder to experience a loss but fail to receive an indemnity payment. By purchasing RICs, however, irrigators would offset the portion of their income risk dependent on adequate water supply. They would be assured of either a sufficient water supply to generate normal crop yields or monetary reimbursement during droughts. This increased income security should create the incentive for irrigators to trade rather than amass excessive water rights since the RIC replaces their need to self-insure.

Theoretically, since the water authority could provide cash pay-offs (indemnities) in lieu of water, it should make less conservative estimates and instead base water right allocations on the most accurate estimates of water supply and demand. Furthermore, since it would be financially obligated to either deliver water or indemnity payments,

Table 1. Indemnity payments (\$100 000) for a rainfall index contract at various coverage levels[a]

		Coverage level				
Cumulative rainfall index	Year	100%	90%	80%	70%	60%
0.21	1982	7.9	7.7	7.4	7.0	6.5
0.55	1997	4.5	3.9	3.1	2.1	0.8
0.58	1980	4.2	3.6	2.8	1.7	0.3
0.71	1987	2.9	2.1	1.1	0	0
0.78	1994	2.2	1.3	0.3	0	0
0.82	1972	1.8	0.9	0	0	0
0.84	1977	1.6	0.7	0	0	0
0.84	1968	1.6	0.7	0	0	0
0.89	1979	1.1	0.1	0	0	0
0.91	1983	0.9	0	0	0	0

[a] Assumes \$1 million in liability.

Source: Australian Bureau of Agriculture and Resource Economics.

the incentives for the water authority to manage the system properly should be significantly improved.

RIC Effects on Water Use

RICs may help reduce total water demand from irrigators by diminishing the hoarding of water rights and on-farm storage of excess water. However, farmers have other strategies that reduce their exposure to water supply risk. The two most significant include investing in more water efficient irrigation technology and changing their crop mix or production methods to those that demand less water. Mahul (2001) finds that under certain conditions risk averse farmers respond to the introduction of weather insurance contracts by increasing their risk exposure. In the case of irrigation, which is a risk-reducing input, his findings imply that some farmers might increase their water supply risk exposure by decreasing their irrigation use, planting more irrigation acres, switching to higher value, higher water use crops, etc. In their more generalized research into the effect insurance has on input use, Chambers & Quiggin (2001) establish a more complicated relationship. They find that providing insurance to farmers causes at least two responses. The pure-risk response agrees with Mahul's (2001) conclusion, i.e. farmers increase their risk exposure. But Chambers & Quiggin (2001) also acknowledge that insurance influences the scale of the farm operation. The scale response can either increase or decrease the level of pure-risk adjustment, depending on how the scale adjustment affects the farmer's risk profile. Thus, with the introduction of RICs, farmers may also increase their operations (planting more acres) and demand more water for irrigation.

Recent research by Carey & Zilberman (2002) finds that the adoption of irrigation technology in the presence of a water market primarily depends on the market price of water. The introduction of RICs would effectively increase the price of water since the irrigator would pay the water market price as well as the insurance premium (either directly or indirectly as part of the price of water). Farmers who sell (buy) their water rights at higher prices have the incentive to invest in water-saving technology. Clearly, the relative price of the insurance is critical to the irrigator's on-farm decisions. If the RIC premium is quite low, farmers may find it less profitable to invest in additional water-saving technology. Therefore, subsidizing RICs may have undesirable effects: less efficient water use on farms and ultimately more demand for water. To avoid this problem, it may be possible to design a more complicated contract with a premium rate that is a function of a farmer's investment in technology that reduces water supply risk (e.g. different crop mixes or improved irrigation technology). However, such a contract would be subject to the same moral hazard problems discussed earlier if monitoring and sound underwriting did not accompany the contract.

The impact of RICs on a farm's conjuctive water system is also unclear. The substitution effect on groundwater use will also depend on the relative price of surface water (which again would reflect the insurance premium). In some instances, the assurance of revenue from the RIC may limit the irrigator's use of groundwater especially if the indemnity payments exceed expected crop revenues minus some accounting for the long-term costs or risks associated with using groundwater (e.g. higher salinity; see Economist (2000)). In other words, with indemnity payments farm income may still increase in times of drought (or at least not decrease) without utilizing any groundwater. In the absence of accounting for groundwater costs, or if the expected net profits are high, farmers may substitute groundwater, produce their crops, and also collect their indemnity payments. More applied water market research is clearly warranted before making any solid conclusion regarding the impact of RICs on water demand.

Implementation

The implementation and efficacy of an RIC depend on sufficient, reliable (tamper-proof) data, accurate pricing, and affordability (Skees, 2000; Martin *et al.*, 2001). While data manipulation may be a concern, individual rainfall station measurements can always be corroborated by comparing them with adjacent stations or with remote-sensing data. Finding lengthy time series data for rainfall and water levels may be a more significant challenge since they are necessary for accurately pricing the contract. Since the RIC premium is a function of the probability of the event, it depends on the number and timing of the years over which the probability is assessed. While lengthy time series of rainfall data certainly exist, more than 50 years of data will still not guarantee an accurate probability assessment. For example, the *El Niño* Southern Oscillation (ENSO) phenomenon could change the probability of the insured events (Podbury *et al.*, 1998). Therefore, it may be necessary to adjust the cost of the insurance when an ENSO event is confirmed, although this would require sufficient lead-time between knowledge of the pending event and the time of selling insurance. Selling multi-year contracts (that cover individual years 3–5 years into the future) prior to ENSO signals may be the better alternative. Multi-year contracts are often used by the reinsurance industry.

In some regions of the world where drought conditions occur frequently, RICs could be prohibitively expensive at higher coverage levels. In these areas, the RIC may only be used to cover infrequent, extreme weather events or low levels of liability (and thus be more affordable). With RICs, like other insurance products, the potential contract holders would choose coverage and liability levels that they could afford. Although they might not be able to afford optimal insurance coverage, if the contract holders are risk averse the RIC would still improve their welfare.

The use of RICs may have the greatest impact in developing countries, where droughts can virtually wipe out a farmer's total farm income as well as their subsistence food. As discussed above, however, the impact of RICs on crop production patterns is not clear. In the case of developing countries, this issue is especially important since RICs could cause a move away from diversified agriculture towards more monocultures. The unique political economies of developing countries pose unique challenges, the full discussion of which is beyond the scope of this paper.

Conclusion

Complex and diverse water systems have been created in many parts of the world to improve water allocation in general and more specifically to manage drought. The difficulties in accurately predicting water supply and demand (both of which vary over time) create inefficiencies in these systems. When temporal trading in water rights is allowed, the risk of drought still decreases the volume of trade since agents are more likely to hoard water rights as a hedging strategy, rather than trade at the margin. This paper presents a partial solution: the incorporation of RICs. Since the water authority could provide cash pay-offs (indemnities) in lieu of water, it should make less conservative estimates and, instead, base water right allocations on the most accurate estimates of water supply and demand. RICs would assure irrigators of either sufficient water supply to generate normal crop yields or monetary reimbursement during droughts. This increased income security should create the incentive for farmers to trade rather than hoard their water rights since the RIC replaces their need to self-insure. However, it should be recognized that the RIC would not be able to reduce the risk of attenuated water rights due to government intervention. In the presence of such risk, the full impact of the RIC would not be realized (i.e. some irrigators would still hold more water rights than necessary as self-insurance against that risk).

It is also important to recognize that at this point it is unclear as to the full effect the introduction of RICs may have on water demand. Irrigators may take on more water supply risk, planting more acres, switching to more high value, high water use crops, etc. The result of this behaviour would be to consume more water, not less, a significant problem for regions such as NSW where water shortages are already severe. More research is clearly needed to accurately assess the impact RICs might have on farm-level water use decisions. It is hoped that future research will incorporate the RIC concept into stochastic models that analyse the management of water supply and demand in uncertain environments. Further research also needs to fully assess the costs and benefits of incorporating RICs into water markets relative to other drought risk management options. In some countries, RICs may not be a feasible alternative. Although the unique conditions of the water management system being considered will ultimately determine the net result insurance will have on drought management, this paper suggests some intriguing possibilities. In the end, few areas of the world can afford to ignore the possibilities of a new approach to drought management.

Notes

1. For example, the water authority would have an incentive to cause a water shortfall if it was the insured (the contract holder) and the potential indemnity payment exceeded expected water market revenues.
2. For a more complete description of the water market situation in NSW see Crase (1999).
3. Murrumbidgee Irrigation, the company that operates the water market in the MIA, posts the allocations on its website (http://www.mirrigation.com.au).
4. The irrigation or 'water' year is typically 1 July–30 June, but drought conditions may cause an earlier deadline (e.g. May).
5. Irrigators who wish to sell their water rights post the quantity of water and price on the water market website (http://www.murrumbidgeewater.com.au).
6. Water users purchase an entitlement and are also charged a fixed price per unit of water based on their entitlement (high security users are charged a higher fee) and water usage (http://www.mirrigation.com.au).
7. During the California drought of 1976–77, net farm income actually increased because the prices of California's specialty crops rose (Howitt & M'Marete, 1989).
8. Different sets of continuous months were tested.
9. The actuarially fair premium = the actuarially fair premium rate × the liability. The actuarially fair premium rate = the expected indemnity/the liability.

References

Arrow, K. J. (1964) The role of securities in the optimal allocation of risk bearing, *Review of Economic Studies*, 31, pp. 91–96.

Bardsley, P., Abey, A. & Davenport, A. (1984) The economics of insuring crops against drought, *Australian Journal of Agricultural Economics*, 28, pp. 1–14.

Beare, S. C., Bell, R. & Fisher, B. S. (1998) Determining the value of water: the role of risk, infrastructure constraints, and ownership, *American Journal of Agricultural Economics*, 80, pp. 916–940.

Carey, J. M. & Zilberman, D. (2002) A model of investment under uncertainty: modern irrigation technology and emerging markets in water, *American Journal of Agricultural Economics*, 84, pp. 171–183.

Chambers, R. G. & Quiggin, J. (2001) Production insurance and input use: an analytical framework, Working Paper 99–03, Department of Agricultural and Resource Economics, University of Maryland.

Crase, L. (1999) Water markets as a vehicle for water reform: the case of New South Wales. Paper presented at 1999 International Symposium on Society and Resource Management, Brisbane, 8 July.

Danzi, E. (1999) Irrigation water reforms, *ABARE Current Issues*, April.

Dudley, N. J. (1988) A single decision-maker approach to irrigation reservoir and farm management decision making, *Water Resources Research*, 24, pp. 633–640.

Dudley, N. J. & Scott, B. W. (1993) Integrating irrigation water demand, supply, and delivery management in a stochastic environment, *Water Resources Research*, 29, pp. 3093–3101.

Economist (2000) The salt that won't run to the sea, 5 February, p. 38.

Howitt, R. E. & M'Marete, M. (1989) The value of groundwater in adapting to drought: lessons from 1976–77, *Coping with Water Scarcity: The Role of Ground Water. Proceedings of the Seventeenth Biennial Conference on Ground Water.* Report 72, California Department of Water Resources.

Howitt, R. E., Watson, W. D. & Nuckton, C. F. (1979) Efficiency and equity in management of agricultural water supplies. University of California, Giannini Foundation of Agricultural Economics, Information Series 79-1.

Jaffee, D. M. & Russell, T. (1997) Catastrophe insurance, capital markets, and uninsurable risks, *Journal of Risk and Insurance*, 64, pp. 205–230.

Mahul, O. (2001) Optimal insurance against climatic experience, *American Journal of Agricultural Economics*, 83, pp. 593–604.

Martin, S. W., Barnett, B. J. & Coble, K. H. (2001) Developing and pricing precipitation insurance, *Journal of Agricultural and Resource Economics*, 26, pp. 261–274.

Miranda, M. J. (1991) Area–yield crop insurance reconsidered, *American Journal of Agricultural Economics*, 73, pp. 233–242.

Musgrave, W. F. (1997) Australia, in: A. Dinar & A. Subramanian (Eds) *Water pricing experiences: an international perspective*, pp. 17–23, World Bank Technical Paper 386, Washington, DC .

Patrick, G. F. (1988) Mallee wheat farmers' demand for crop and rainfall insurance, *Australian Journal of Agricultural Economics*, 32, pp. 37–49.

Podbury, T., Sheales, T. C., Hussain, I. & Fisher, B. S. (1998) Use of El Nino climate forecasts in Australia, *American Journal of Agricultural Economics*, 80, pp. 1096–1101.

Quiggen, J. C. (1986) A note on the viability of rainfall insurance, *Australian Journal of Agricultural Economics*, 30, pp. 63–69.

Sanderson, F. (1943) A specific risk scheme for wheat crop insurance, *Journal of Farm Economics*, 25, pp. 759–776.

Skees, J. R. (2000) A role for capital markets in natural disasters: a piece of the food security puzzle, *Food Policy*, 25, pp. 365–378.

Skees, J. R. & Barnett, B. J. (1999) Conceptual and practical considerations for sharing catastrophic/systemic risks, *Review of Agricultural Economics*, 21, pp. 424–441.

Topp, V. & McClintock, A. (1998) Efficiency and equity in water markets. Paper presented at the Australian National Committee on Irrigation and Drainage (ANCID) 1998 conference, Sale, Victoria, 2–5 August (Australian Bureau of Agriculture and Resource Economics (ABARE) Conference Paper 98.19).

Turvey, C. G. (2001) Weather derivatives for specific event risks in agriculture, *Review of Agricultural Economics*, 23, pp. 333–351.

INDEX

Page numbers in **bold** refer to figures. Page numbers in *italics* refer to tables.